Art + DIY Electronics

Leonardo

Seán Cubitt, Editor-in-Chief

Social Media Archeology and Poetics, edited by Judy Malloy, 2016

Practicable: From Participation to Interaction in Contemporary Art, edited by Samuel Bianchini and Erik Verhagen, 2016

Machine Art in the Twentieth Century, Andreas Broeckmann, 2016

Here/There: Telepresence, Touch, and Art at the Interface, Kris Paulsen, 2017

Voicetracks: Attuning to Voice in Media and the Arts, Norie Neumark, 2017

Ecstatic Worlds: Media, Utopias, Ecologies, Janine Marchessault, 2017

Interface as Utopia: The Media Art and Activism of Fred Forest, Michael F. Leruth, 2017

Making Sense: Cognition, Computing, Art, and Embodiment, Simon Penny, 2017

Weather as Medium: Toward a Meteorological Art, Janine Randerson, 2018

Laboratory Lifestyles: The Construction of Scientific Fictions, edited by Sandra Kaji-O'Grady, Chris L. Smith, and Russell Hughes, 2018

Invisible Colors: The Arts of the Atomic Age, Gabrielle Decamous, 2018

Virtual Menageries: Animals as Mediators in Network Cultures, Jody Berland, 2019

From Fingers to Digits: An Artificial Aesthetic, Ernest Edmonds and Margaret A. Boden, 2019

MATERIAL WITNESS: Media, Forensics, Evidence, Susan Schuppli, 2020

Tactics of Interfacing: Encoding Affect in Art and Technology, Ksenia Fedorova, 2020

Giving Bodies Back to Data: Image-Makers, Bricolage, and Reinvention in Magnetic Resonance Technology, Silvia Casini, 2021

A Biography of the Pixel, Alvy Ray Smith, 2021

Living Books: Experiments in the Posthumanities, Janneke Adema, 2021

Art in the Age of Learning Machines, Sofian Audry, 2021

Design in Motion: Film Experiments at the Bauhaus, Laura Frahm, 2022

Northern Sparks: Innovation, Technology and the Arts in Canada from Expo '67 to the Internet Age, Michael Century, 2022

The Artwork as a Living System: Christa Sommerer and Laurent Mignonneau 1991–2022, edited by Karin Ohlenschläger, Peter Weibel, and Alfred Weidinger, 2023

Computational Formalism: Art History and Machine Learning, Amanda Wasielewski, 2023.

Picture Research: The Work of Intermediation from Pre-Photography to Post-Digitization, Nina Lager Vestberg, 2023

Art+DIY Electronics, Garnet Hertz, 2023

See http://mitpress.mit.edu for a complete list of titles in this series.

Art + DIY Electronics

Garnet Hertz

The MIT Press
Cambridge, Massachusetts
London, England

The MIT Press would like to thank the anonymous peer reviewers who provided comments on drafts of this book. The generous work of academic experts is essential for establishing the authority and quality of our publications. We acknowledge with gratitude the contributions of these otherwise uncredited readers.

This book was set in Stone Serif by Westchester Publishing Services. Printed and bound in the United States of America.

Library of Congress Cataloging-in-Publication Data

Names: Hertz, Garnet, 1973– author.
Title: Art + DIY electronics / Garnet Hertz.
Other titles: Art and DIY electronics
Description: Cambridge, Massachusetts : The MIT Press, [2023] | Includes bibliographical references and index.
Identifiers: LCCN 2022026917 (print) | LCCN 2022026918 (ebook) | ISBN 9780262044936 (paperback) | ISBN 9780262361583 (epub) | ISBN 9780262361576 (pdf)
Subjects: LCSH: Art and electronics. | Maker movement. | Arts—Experimental methods. | Technology—Social aspects.
Classification: LCC N72.E53 H47 2023 (print) | LCC N72.E53 (ebook) | DDC 702.8—dc23/eng/20220727
LC record available at https://lccn.loc.gov/2022026917
LC ebook record available at https://lccn.loc.gov/2022026918

10 9 8 7 6 5 4 3 2 1

For Amelia, Emma, Max, Kohl, and Arlo

Contents

Series Foreword

Leonardo/The International Society for the Arts, Sciences and Technology fosters transformation at the nexus of art, science, and technology because complex problems require creative solutions. The Leonardo Book Series shares these aims of artistic and scientific experimentation, and publishes books to define problems and discover solutions, to critique old knowledge and create the new.

In the early twentieth century, the arts and sciences seemed to interact instinctively. Modern art and modern poetry were automatically associated with relativity and quantum physics, as if the two were expressions of a single Zeitgeist. At the end of the Second World War, once again it seemed perfectly clear that avant-garde artists, architects, and social planners would join cyberneticists and information theorists to address the problems of the new world order and to create new ways of depicting and understanding its complexity through shared experiences of elegance and experiment. Throughout the twentieth century, the modern constantly mixed art and science.

In the twenty-first century, though, we are no longer modern but contemporary, and now the wedge between art and science that C. P. Snow saw emerging in the 1950s has turned into a culture war. Governments prefer science to arts education, yet stand accused of ignoring or manipulating science. The arts struggle to justify themselves in terms of economic or communicative efficiency that devalues their highest aspirations. And yet never before have artists, scientists, and technologists worked together so closely to create individual and collective works of cultural power and intellectual grace. *Leonardo* looks beyond predicting dangers and challenges, beyond even planning for the unpredictable. The series publishes books that are both timely and of enduring value—books that address the perils of our time, while also exploring new forms of beauty and understanding.

Seán Cubitt, Editor-in-Chief, Leonardo Book Series

Erica Hruby, Editorial Director, Leonardo/ISAST

Acknowledgments

Thanks to Billy Klüver and Mark Pauline for talking to me on the telephone in 1995—I was a kid calling out of the blue from Saskatchewan, and you let me in to your worlds through phone interviews that shifted the path of my life. Thanks to Douglas Repetto who started the informal artist / hacker organization Dorkbot in NYC in 2000, and Karen Marcelo in San Francisco who inspired me to organize Dorkbot in Southern California in 2004—we honestly saw the birth of the entire "maker" identity from underneath us in 2005. In many ways, the construction of this book can be seen as an effort to link Klüver, Pauline and Dorkbot together and to articulate why creating "alt tech" as artwork still holds significant cultural value—and how it helps us understand the rich complexity of being human.

Many thanks to the many artists that talked to me or were interviewed as a part of this research project. Thanks to Diana Burgoyne, Natalie Jeremijenko, Nancy Paterson, Darsha Hewitt, Rich Pell, Thomas Thwaites, Karen Marcelo, Mark Pauline, Reed Ghazala, Norm White, Jeremy Bailey, Dmitry Morozov, Rob Cruickshank, Michael Shiloh, Shih Chieh Huang, Tom Jennings, and Nell Tenhaaf for talking to me. Also, thanks to the archivists, technicians, and gallery staff that pulled out items from their remarkable archives to help me get a better understanding of these projects. Special thanks to Rhys Edwards and the talented staff at the Surrey Art Gallery, Jean Gagnon and the helpful folks at the Cinémathèque québécoise in Montréal, the archivists at the National Gallery of Canada, and Julia Novakovic at the Brian Sutton-Smith Library and Archives of Play at the Strong Museum. Thanks to Molly Steenson and Robin Oppenheimer for their information on Experiments in Art and Technology, and thanks to Anthony Dunne and Fiona Raby for access to their project documentation.

Thank you to the individuals that provided valuable feedback, research, and editing help, including Matt Harkness, Liam Bannon, Edward Shanken, and Sarah Cook. Special thanks to Andrew Schrock and Amelia Guimarin, who patiently and skillfully went through and carefully edited the entire manuscript with me. Thanks to my undergraduate students at Emily Carr University that provided feedback, including Nikki Reyes, Fergus Clarke, Ember Johnstone, Felicia Sugiarta, Stephanie Ford, Kala Brownlee, Yvonne Zhao, Rachel Kim, Shriesha Sasank and many others. Thanks to the helpful reviewers that heard my call for help on Facebook and gave me feedback: Deepa Butoliya, Rob Duarte, David Forbes, Danny Spitzburg, Tina Rivers Ryan, Michelle Kasprzak, Rafael Lozano-Hemmer, Wladimiro Woyno, Gabriel Menotti, Karen Marcelo, Jennifer Willet, Rob Cruickshank, and David Li. Also, thanks to the graduate students I have worked with at various institutions that have provided me with inspiration, conversation, and ideas that have directly impacted this book, including Emit Snake-Beings (University of Waikato), Yana Boeva (York University), Andrew Quitmeyer (Georgia Tech), Matt Harkness (Emily Carr University), Minka Stoyanova (City University of Hong Kong), Regina Sipos (Technischen Universität Berlin) and Kaiwen Yang (Emily Carr University). A massive thanks to Kaiwen Yang for helping with many aspects of the final production of this project, including formatting bibliographical sources and securing all the image rights for this book.

My wonderful and caring colleagues at Emily Carr University have helped contribute to improving the work environment and I am thankful for them: Leanne Rooney, Keith Doyle, Hélène Day Fraser, Bobbi Kozinuk, Kate Armstrong, Valérie Walker, Rita Wong, Amber Frid-Jiminez, Richard Hill, Maria Lantin, Alan Goldman, Haig Armen, Katherine Gillieson, and many others. Thanks to Noel Rubin for helping me keep sane during the end of the writing process at the Vancouver Public Library.

This book is made possible through the gracious support of the Canada Research Chair Program, the Social Sciences and Humanities Research Council of Canada, and Emily Carr University of Art and Design—this book would never have been completed were it not for you. I am incredibly grateful for the opportunity to work on this project. Thank you. Support for foundational aspects of this book was also provided through many institutions at the University of California, Irvine, including the Institute for Software Research, the Laboratory for Ubiquitous Computing and Interaction, and the California Institute for Telecommunications and Information Technology. Thank you to Mark Poster (who is missed), Peter Krapp, Paul Dourish, Gillian Hayes, Don Patterson, Geof Bowker, Cecile Whiting, and Walt Scacchi for their support in foundational work that led to this book. You folks are exceptional.

Lastly, thank you to my immediate family for loving me, caring for me, and laughing with me during the development of this project: you are the world to me.

Beginnings An Introduction to the DIY Mindset

Junk Piles and Rural Lessons from Clemenceau, Saskatchewan

My interest in do-it-yourself culture is influenced by my childhood on a farm in Clemenceau, Saskatchewan (figure 0.1). Rural technologies, often fixed or built out of necessity, sparked my love for innovations: fixing a throttle linkage with a coat hanger, building a go-kart out of a tricycle, and trying to weld together hybridized farm equipment to save time and make money. People have to "make do" in rural environments without professional help or equipment. As a result, the built objects are often quirky, kludged, and nonstandard.

However, farms in Saskatchewan are far from utopian. Immigrants to Canada around the turn of the twentieth century were promised wide-open, bucolic farmland. Instead, they found a rugged environment with harsh winters, a fog of mosquitoes during the short summers, and great distances between settlements. Although vast, the territory was not empty. The settlers who came to claim the land often did so through brutal force at the expense of indigenous communities that had lived there for thousands of years. Amplified by the weather, natural terrain, and decades of systematic neglect, the infrastructural gaps in Saskatchewan are still considerable—especially in remote communities and among the First Nations. Surviving there requires resourceful and creative use of available materials, especially for those with less money and lower sociopolitical status. In other words, what we call "DIY" (do-it-yourself) is regularly a necessity to navigating and bridging gaps in infrastructure.

Clemenceau taught me a counterpoint to naive technological optimism. Rural Canada was idyllic, but also a place of technical need and conflict coupled with a lack of resources. My childhood was not spent watching television—we only had two channels—but playing in abandoned cars,

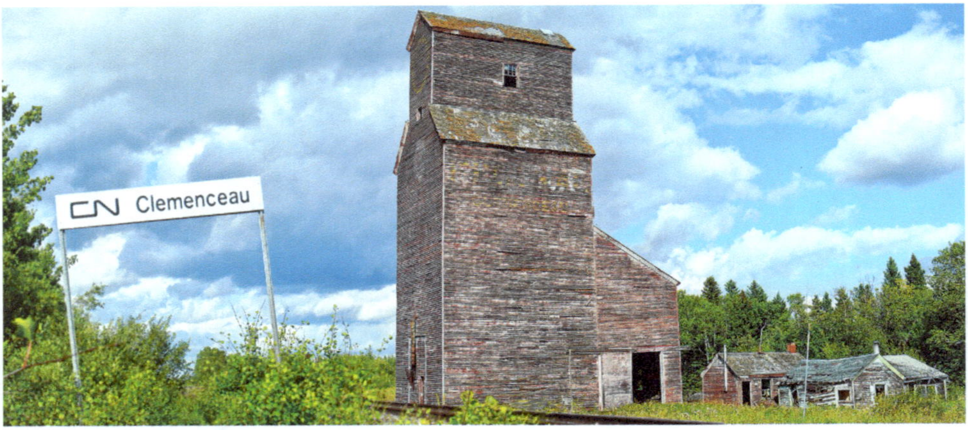

Figure 0.1
A lone grain elevator stands in Clemenceau, Saskatchewan, a town of fifteen people according to the 2011 Canadian Census. Photo Garnet Hertz.

kludging together forts, and assembling go-karts (figure 0.2). As I got older, memories of Clemenceau tempered my enthusiasm for the transcendental promises of digital technologies. Richard Barbrook and Andy Cameron called this promise "The Californian Ideology"—the idea that information technologies were economically beneficial and emancipatory for everyone. This language is reflected in publications like *Mondo 2000*, *Wired* magazine, and concepts like "new media." However, the Californian Ideology of technology being liberatory lacked a connection to the world that I understood in rural Saskatchewan. Rural life demanded that I grapple with the malfunction and breakdown of technology. The things I built were haphazard, awkward, and even junky. Rather than causing me to bemoan the obsolescence of technologies, my childhood showed me that working with old and seemingly worthless objects could unlock opportunities for creative freedom.[1]

With the advent of the Internet, I found myself fascinated by my ability to communicate with others in remote locations. In 1994, my worldview was changed by talking over Usenet newsgroups and accessing information on distant FTP servers around the world.[2] I was quickly swept up in communities of artists and other people hacking and playing around with technology, while also remaining skeptical of the language used to describe cyberspace, virtual reality, and new media. I became fascinated with a budding DIY culture that predated the contemporary fascination with "making." As I learned about this undercurrent of experimentation, the DIY

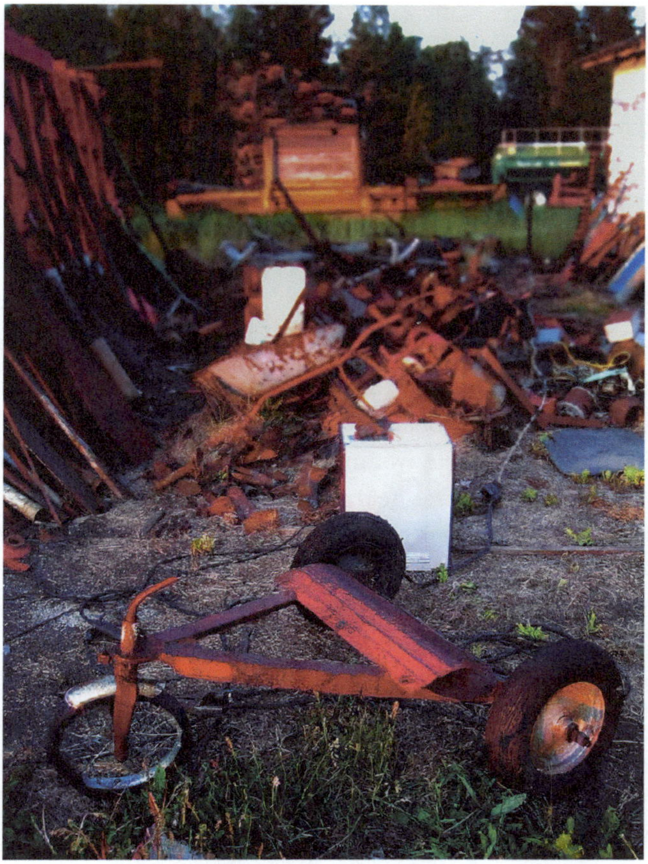

Figure 0.2
Remnants of the go-kart I built from scrap materials with my family in Clemenceau, Saskatchewan. Photo Garnet Hertz.

artists you'll read about in this book became my heroes. Mark Pauline of Survival Research Labs built large-scale creative robotic devices.[3] Groups like Experiments in Art and Technology brought together artists and engineers to build experimental projects. Nancy Paterson took telerobotic devices to an entirely new level. Others hacked motors and electronics in creative and unexpected ways.[4] Sitting in my parent's basement on a 486DX computer with a dial-up modem, I started tracing a creative community of Clemenceau-style DIY artists across the globe. To borrow Foucault's term, this distributed community of DIY-style, low-budget hackers was a "heterotopia" or flipside of new media—almost an underbelly or counterculture of technologists that generally called bullshit on the unattainable promise of a utopia in cyberspace.[5]

Building a DIY *Telegarden*

As I slowly learned about the history of electronic art, I noticed a largely untold story unfolding about the intersection of DIY approaches to technology and artistic practice. Published books on art and technology, like Frank Popper's influential *Art of the Electronic Age* from 1993, present only work that looks professionally fabricated, polished, and complete.[6] The exhibitions, festivals, and studios I visited told a completely different story, however. I saw that the vast majority of artists working with technology were cobbling together projects with their own makeshift tools and methods. Often, I and all the artists I knew in the field struggled just to keep our projects running. In other words, the desire to get art and technology practices taken seriously by the discipline of art often swept the messy DIY processes with technology under the rug. This book helps correct this erasure by outlining themes that undergird DIY practices in contemporary art and experimental design fields and among the wider public.

I remember well my own first attempt to build a piece of robotic art. At the time, I was familiar with Ken Goldberg's *Telegarden* project at the University of Southern California, which clearly demonstrated the potential of using the Internet for robotics.[7] The *Telegarden*, which was launched in June 1995, consisted of a large Adept-1 industrial robotic arm placed in the center of a circular garden container filled with soil. The arm was equipped with a camera connected to the Internet, and custom software enabled remote users to move the arm to plant and water seedlings. Remote participants could see and care for their growing projects. For its time, the *Telegarden* was technologically advanced and futuristic. Viewing the live webcam and controlling the system from a dial-up modem over 3,000 kilometers away in Saskatchewan was uncanny. However, the general framing of the system and curation of the operating project struck me as missing the most interesting part of the work: showing a physical manifestation of interpersonal conflict online.[8]

The team described the *Telegarden* as a "'post-nomadic' community where survival favors those who work together."[9] However, as the global community collaboratively gardened and cared for plants, rogue users occasionally sabotaged the team's vision of a peaceful futuristic landscape. They intentionally overwatered plants and extended the end of the arm into the dirt to rip up the soil and raze the vegetation. The *Telegarden* team moderated out subversive activity and users in the environment. From the perspective of the project, this made sense; it was designed to be a *garden*, not a sandbox or mud pit. However, I found this remote physical sabotage the most interesting social component of the work—or at least it seemed common sense to the farm kid in me that this would happen. The idea of remote users operating a robot arm to rip up dirt was a potent counterpoint to a more professional image, one that would have likely preferred a more functional, frictionless, and pleasant cooperation through telepresence. In response, in 1995 I kludged together a telerobotic device that amplified the uncontrolled and open-ended components that were downplayed in the *Telegarden*. My idea was to build an Internet-controlled machine that enabled remote users to make a mess and draw on the floor, without any rules.

I began work on *Interface* (figure 0.3) by cobbling together parts of an old remote-controlled car, welding and bolting together prototypes for my Internet-controlled drawing machine. I connected the device to a serial motor controller through my Macintosh Performa 5200CD. An antiquated video camera I got for free dangled above the machine, connected by rope and duct tape. Using kludged together software, I connected a video capture card to a web-based interface. This hacked-together system operated in a friend's warehouse studio space and the software frequently crashed. Since this was long before the Arduino or webcam systems were available commercially, I had to build it in a DIY way. The servo arms would fall out, and its gravity-fed drawing mechanism would regularly jam. As a result, the *Interface* project operated reliably for only a handful of hours in April 1996.[10] Still, I was inspired by my experience of building a telerobot for a couple hundred dollars. This Clemenceau-esque *Telegarden* set me on the path of developing projects and researching creative technologies, especially the innovative works that artists, hackers, and experimental designers were building under the radar.

My mindset of exploring the dirtier DIY underside of telerobotics were echoed by Rafael Lozano-Hemmer in a short article, "Perverting Technological Correctness" in 1996.[11] Lozano-Hemmer suggested that electronic art projects should critically question the role of technology in culture. He did not believe that users should merely envision a harmonious future

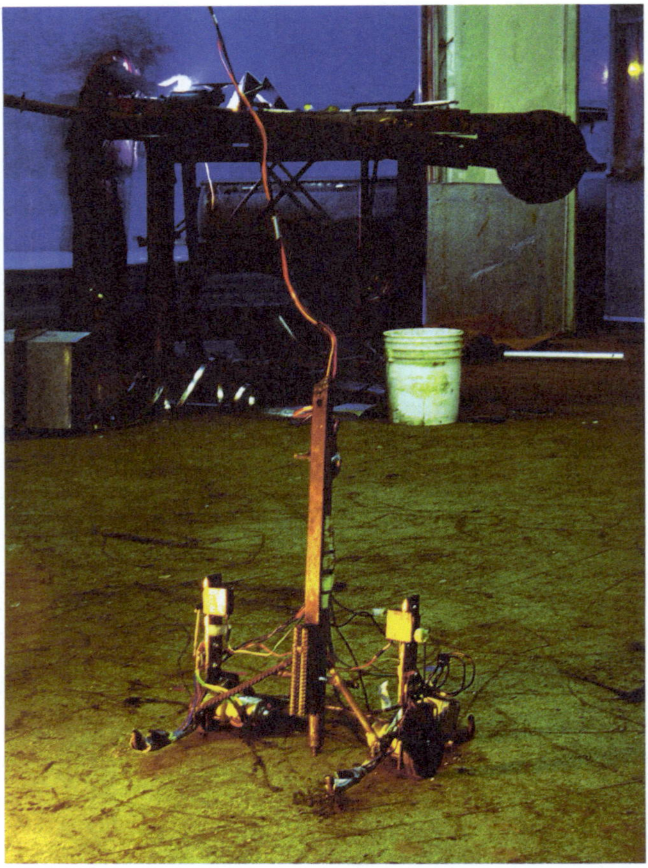

Figure 0.3
Garnet Hertz, the *Interface* telerobotic project, April 1996. Photo Garnet Hertz.

or marvel at technological capabilities such as speed or resolution. Rather, the projects Lozano-Hemmer promoted were intentionally difficult to use, uncomfortable, or limited—but they were done so with care, and invited intervention, criticism, and humor. In many senses, Lozano-Hemmer's *perverting of technical correctness* forecast the mindset of *critical design* before Anthony Dunne coined the term in 1999.[12] By "perverting" technological correctness, he swapped out the usual optimism and technological determinism, creating something closer to the work of Survival Research Labs, the people sabotaging the *Telegarden*, and my Internet-controlled drawing machine.

ISEA as an Undercover DIY Festival

Artists often wrestle with the media of their practice. This is true whether they struggle with the constraints of paint, printmaking, metal, stone, or photography. Electronic and media art practice is no exception, because it has always oscillated between *vast creative promises* of new technologies and the *overwhelming complexity* of new technologies. The more experimental the media is, the more it pushes artists to find specific technical solutions. These difficulties were especially evident at the 1997 International Symposium of Electronic Art (ISEA) in Chicago. Many pieces were not functioning, and galleries were full of artists frantically trying to debug motors, circuits, or software. The experience of being there was more like a DIY-style "hackathon" than the refined and graceful photos I had seen in Frank Popper's *Art of the Electronic Age.*[13] Artworks by the Center for Metahuman Exploration[14] and Myron Krueger[15] were technically complex and consistently operated. However, the most compelling installations to me were technically tentative.

Eduardo Kac made an aluminum and glass robot titled *A-positive* that had a flame powered by his blood. During the exhibition, a doctor inserted a needle into Kac's arm, triggering the contraption to start collecting blood. It was strange and captivating, but it only actually operated according to plan for a few seconds before the ignition chamber overflowed with blood and extinguished the fire.[16] In an adjacent room, Philippe Demers and Bill Vorn were tuning a control system to tame their scraping and flailing pneumatic robots well after the show had opened.[17] Even the most technologically advanced pieces—like the virtual reality system presented by Char Davies in her talk at the event—looked and felt like an awkward tangle of cables.[18]

In hindsight, the new media exhibition could not fully sweep the DIY process of *designing* electronic art under the carpet. Rather, I saw the technical struggles that the artists had gone through as leaving an indelible mark. The visibility of their bespoke technical processes became an inescapable part of the artwork. Simultaneously dismayed and encouraged, I left ISEA97 disenchanted by how most of the work did not live up to the promise and potential of a futuristic "new media." Yet, I was also intrigued that my humble Internet-controlled drawing machine might have fit in alongside flailing robots and blood-extraction performance art. The creativity of the *DIY process* was fascinating. Although exhibiting new media was the goal at

ISEA97, the exhibition seemed more about the DIY practice undergirding the still-nascent idea of what new media might be. This book gives these types of projects a name—DIY electronic art—and contrasts their gritty realities with what has come to be called "making." It strives to articulate a language and rubric of DIY culture.

How the Maker Movement Was Born

I titled this book "DIY" to draw attention to an unacknowledged political undercurrent of art and design that transcends disciplinary boundaries and rejects commercialism as a core focus. I see DIY in contrast to making, which emerged as a concept when Dale Dougherty coined the term in 2005 when he published a new magazine called *Make*. The do-it-yourself attitude of making or maker culture surged around 2012, especially in North American contexts.[19] The category of "maker" greatly spread the concept of DIY tech culture, but in the process of popularization, it lost the core "punk tech" attributes of DIY art that excited me.

Part of my frustration with making was influenced by my having a front row seat to the birth of the idea of what a *maker* is. I moved to the Los Angeles area in 2003 and started Dorkbot SoCal—a Southern California branch of a grassroots organization started by Douglas Repetto in 2000. Repetto, who had moved to New York City for a job as a computer music professor at Columbia University, started the Dorkbot organization as a way to meet other people doing experimental projects with electronics. He founded the organization in New York as a demo event where three people would have twenty minutes each to show their work under the slogan of "people doing strange things with electricity." Soon, its popularity inspired him to open up the organization to people in other cities. My friend Karen Marcelo (of Survival Research Labs) had started Dorkbot SF in San Francisco in June 2002, so I started a similar group in Southern California called Dorkbot SoCal in April of 2004.

Marcelo and I had front row seats to the birth of what came to be known as the maker movement. We discussed the *Make* project with its originators before it began—Mark Frauenfelder went to my events in Los Angeles and Dale Dougherty would go to Karen's in San Francisco. Our informal Dorkbot events laid out a particular framework, community, and mood that *Make* magazine would later capitalize on, with many of our presenters, and the technologies they showed forming the original content of *Make*. This was welcome, although *Make* sanitized and depoliticized many DIY-style experiments for a mass audience. In my estimation, the edgier and more artistically inclined projects shared through Dorkbot were considerably more conceptually interesting than how they were rendered under the new banner of making.

The DIY Mindset

This book tracks and traces the contours of an alternate lineage to the maker movement—one that I refer to as "DIY" or "DIY electronic art." It intentionally is not a movement or revolution, but more like a trend or personality. A "DIY mindset" of sorts. While DIY and making have similarities, I argue that DIY is distinct from making—and the bulk of this book carves out the attributes of what DIY practice *is*, how it works, and what its benefits are. DIY is embodied in and attentive to process, uses authenticity and materiality for political ends, and embraces a fuller and more complex notion of the amateur that rejects management. DIY practices also work especially well when resources are limited, as an educational platform, or when people need to find identity and meaning in their lives. DIY practices also are essential for countercultures—to build identities and resources, or as tools of protest. Yes, the maker label has been useful and insightful as an interdisciplinary platform since its creation in 2005—Dale Dougherty should be strongly commended for seeing and giving a name to this cross-disciplinary trend. However, it does nothing to show how we got here and is relatively aimless, scrambling and unable to chart a path forward. Accordingly, my goal in this book is to unearth this vibrant history to show how DIY has *always been* and *continues to be* part of contemporary artistic practice. In some senses, the core thing wrong with the maker movement is that it never acknowledged that it is at least a century old in the field of art. It's not a movement, it's a mindset.

Maker culture is not as euphoric or cool as it used to be—and it is completely fine if "DIY culture," "punk technologies," or something else replace it as a concept. Dougherty had envisioned *Make* as a "Martha Stewart for Geeks" when first proposing the publication, which completely misses the point that people design and build things for more than just recreational or hobbyist goals.[20] A mutual threat to maker and electronic art circles is also technophilia—the fascination with new technologies without context or historical perspective.[21] The lack of vision in the maker movement was thrown into stark relief on June 9, 2019 when Maker Media—who published *Make* magazine and organized hundreds of Maker Faires internationally—announced that the company had ceased operations. Dougherty cited a drop in industry sponsorship as the key to the company's demise. However, I believe Maker Media's failure stemmed from how it failed to engage with the world beyond leisure technological gadgetry. It primarily survived by surfing two waves of hype: the Arduino wave around 2010 and the 3D printing wave that crested in 2012. However, the path forward requires us to look beyond making to reclaim the histories and themes that can propel a new community forward. We need to resuscitate DIY to save making from itself.

My reasons for rejecting making in favor of DIY should now be clear. The inescapable problem with making is that it trades out the cultural, artistic, and controversial components of DIY culture for a more family-friendly format for a mass audience palate. The danger of the maker movement is that it approaches engagement with technology through the totalizing lens of *Make* magazine's lifestyle tech brand, neglecting the many older strands of electronic art that it is inspired by. "Maker culture" has been useful as a term to bring a diverse range of people together, but it is relatively incomplete because it is generally void of the complexity, history, and thorny political possibilities within art and design. In response, this book articulates one of the core strands left out of this Martha-Stewartization of DIY into maker culture that reaches back nearly a hundred years: DIY electronic art. This book naturally shares an affinity with other technologically oriented countercultures, like the history of hacker culture, the free software movement, experimental and noncommercial designers, and the FabLab movement. However, my emphasis is primarily on contemporary artistic practice, a theme sorely absent from official histories.

Looking back at these influences on the contemporary idea of making helps us see their historical influences and the morphology of the concept over time. DIY electronic art revives these lessons from the past to point a new path forward for artistic practice and design work. The maker movement's technological fetishization will disappear as fast as the waning hype around 3D printing: nobody is really excited by 3D printing anymore anyhow. By contrast, this book's enduring history centers on experimental art and design practices, themes that will continue to nourish future generations disappointed at how "making" has aged. This book shows that DIY electronic art has a history that is worthy of defending when a label like maker slowly fades from popular culture.

Electronic Art as DIY Practice and Theory

There have been few efforts to write a history of electronic art from a DIY perspective. This book strives to fill that void by outlining a lexicon of electronic art through a DIY lens, an exercise that traverses contemporary art practices, experimental design, and science and technology studies (STS). While I am inspired by the energy of the maker movement, I see it as a short-term trend in comparison to design, art, and human–computer interaction. "Maker" as a term is a bit of a fad. In hindsight, the maker movement was often stuck between two extremes when looking at technology: it either saw technology as a method of weekend entertainment, or as a type of open source religion. It fixated more on new technologies than on processes. However, it mobilized communities around creative technological work in highly innovative ways across the globe. Although both artists and makers have connections to commercial enterprise, their motivations are also primarily personal and exploratory: artists, experimental designers, and makers are often motivated by goals other than making money.[22]

Maker culture draws strength from playfulness and a willingness to experiment. I invite any maker folks reading this to seriously consider themselves "artists." Although the art world does not have a prominent section devoted to technological exploration, it certainly has a fascinating history of artist-hackers that reaches back a long, long time. Makers often miss this rich history of technical creativity, and I have written this book to help fill that gap. Artists are concerned with a vibrant range of topics—the intricacies of perception, politics, and technologies, or the exploration of what it means to be human—that honestly enrich the contemporary electronic maker community. Accordingly, by writing the history of electronic art as a form of DIY practice, I hope to chart a historically informed and engaged path forward for makers.

Scope of Book

This book starts by providing a thematic and critical overview of electronic art, a field that is underhistoricized and undertheorized. I see electronic art as involving electromechanical practices—a distinction between it and screen-oriented or audio-focused work like experimental music, video art, software-oriented art, net art, and video game art. Accordingly, I also do not include technologies or practices outside of the art context, such as hobby radio work, or related practices like "phreaking" (phone hacking).

Instead, this book focuses on projects circulating in the sphere of art that use electronics as their primary medium. I also take the liberty of including some people who think of themselves as designers instead of artists—I propose a thoroughly permeable and malleable line between the fields of art and design. Organizationally, however, this book responds to the tendency for media art and art history texts to overindex or overtheorize projects. Overindexing lists out as many examples based on topical themes as possible, without excavating their historical context or the artists' creative practices.[23] Overtheorizing typically props up projects around a European critical theory. My issue is less with having too much theory, and more that this approach tends to be a top-down approach distanced from the highly individualized, quirky, and personal "bottom-up" motivations of artists. The artwork demands it.

Instead of overindexing and overtheorizing, this book takes a lexical approach. It is less of a comprehensive art history of DIY electronics, and more of an exploration of notable case studies and themes. Through extensive interviews and field research, I have listened to the intentions of the creators. By analyzing electronic art, I reveal the technical details of the work and work to communicate them with clear and straightforward language.[24] Organizationally, in each chapter I have pulled out a number of key concepts as a core attribute of DIY practice in general, with the aim of creating a general-purpose lexicon of DIY. This lexical approach draws more from design theory and science and technology studies than art history. In many ways, this book strives to be a clearly written design and STS analysis of DIY technical production. I use examples from art and critical design that can be read and understood by artists, designers, and readers entirely new to electronic art. The relatively short chapters have been crafted with students and educational environments in mind. Think of them as self-contained vignettes that can be assigned as readings in art, media studies, technology studies, cultural studies, and digital media programs.

It should also be noted that this book does not spend much time delving into either definition of art or design. In fact, this book intentionally blends these two categories together. This is partially because these practitioners often live and work between the formal world of contemporary art and the diverse world of design—the majority of these individuals refer to themselves as both "artists" and "designers" depending on the professional context. As a result this book does not work to articulate a new definition of art. Instead, it articulates the characteristics of DIY electronic practice with individuals that intersect with contemporary art. Clement Greenberg, Rosalind Krause, Roland Barthes, Meyer Schapiro, and other art historians have been put aside in favor of science and technology scholars, design academics, researchers of amateur culture, and theorists of innovation.

The direction taken in this book tries to balance an inductive and technical analysis that will please hackers with a scholarly rigor that will please technology academics. By exploring projects in depth and with historical context, I work to carve out categories that connect projects with concepts that help explain why electronic art is distinct and notable. Through innovative case studies and articulated implications for design, the text charts a new path for creative electronic practice to move beyond making unconventional technologies and into a critically and historically engaged studio practice.

Foundations and Terms: Book Layout

As a lexicon of DIY electronic art, this book is organized in two core parts: a foundational part that provides a historical overview of electronic art and the term "DIY," and a second part with detailed case studies of artist-created technologies that are used to explore additional concepts of relevance to DIY culture and electronic art. The first part has two chapters that each explore a definition: *electronic art* and *DIY*. Chapter 0.1, "A History of Electronic Art in the Twentieth Century," charts a history of electronic art and proposes that it is still a useful contemporary category. Chapter 0.2, "A Definition of DIY: Do + It + Yourself," lays a theoretical groundwork for what "do-it-yourself" is: what is "done," what "it" is done to, and "who" does it. The frameworks from these two chapters lay a theoretical foundation for DIY electronic art for the case studies in the rest of the book.

The second part—which makes up the bulk of the manuscript—presents thirteen chapter-length case studies of art and design projects from a DIY perspective. Each chapter makes an effort to look at the actual material and technical specifics of the projects. Throughout my writing, I am motivated by the belief that artists are designers of technology, so the technical details matter. Artists are inventors. To art historians and art critics, this may come across as getting bogged down in technical details. My response to this suggestion is that many historians and critics would do better scholarship if they understood the technical processes and material realities of building artwork. Artists do more than generate ideas—they are clearly also designers and hackers who engineer technologies in novel ways. Outlining the inner workings of historically significant projects helps illustrate how DIY practices have produced raw innovation in technical culture at large. These artists are often significantly more interesting and innovative than their commercial startup counterparts. I invite you to explore their curious stories in the pages ahead.

Each chapter pulls out a number of themes behind the work—a few core concepts and ideas—that helps us understand how DIY practices work in general. These thirteen chapters illuminate both well-known and relatively undiscovered works and incrementally carve out a robust lexicon of DIY technology practices and culture. Or at least I hope it is the start of a lexicon. I have paid close attention in order to select a diverse range of works in this field, but many examples chosen as sigificant case studies are drawn from projects I have observed in person or from artists and designers I have personally met or talked to. This skews many examples to be limited to my global footprint in North America and to generally be people like me. It is my hope that this is viewed as only a start of an exploration of this style of work within the field of art discourse and science and technology studies—and not as a concluding summary of the scene. In a way, these are examples from my personal and semilocal scene.

The thirteen chapters are brought together in five thematic clusters: frugality, exploration, identity, disobedience, and selling out. The first theme—frugality—explores projects where economic and material constraints are an essential part of artistic do-it-yourself practice. After an introduction to the theme of frugality (theme 1), three case studies fill out the first theme: chapter 1.1, "Frugality and the *Demanufacturing Machine*: Zombie Technology, Bricolage, and Hype Cycles," describes the first machine built by Survival Research Laboratories in 1979; chapter 1.2, "Frugality and *Telephonic Arm-Wrestling*: Jugaad, Finances, and Function," explores a project built by Doug Back and Norm White in 1986; and chapter 1.3, "Frugality and the

Toaster Project: Technical Disorientation, Device Paradigms, and Highlowness," features a self-produced electric toaster built by Thomas Thwaites in 2009. Taken together, these chapters that explore constraint in DIY see it as three things: a practice where the old and obsolete are given new life, a practice where frugal innovation is done out of necessity due to a lack of resources, and a practice where constraints can be chosen intentionally.

The next cluster of chapters (theme 2) explores the theme of taking technologies apart and exploring them—doing something yourself in order to get a better understanding of how something works. The three chapters along this theme also dig into artistic case studies and work to articulate a few concepts per chapter. Chapter 2.1, "Exploration and the *Incantor*: Bending Circuits, Depunctualization, and Unblackboxing," looks at the practice of circuit bending through the work of Reed Ghazala. In this case, technologies are broken open and hacked primarily for the purpose of creatively exploring unknown technologies. Chapter 2.2, "Exploration and *Wire Figures*: Technologies, Interactivity, and Radio Shack Cybernetics," takes the electronic artwork of Diana Burgoyne as a starting point to discuss DIY practice motivated by a desire to make technology more legible. The cluster on revealing technologies wraps up with chapter 2.3, titled "Exploration and *20 Oscillators in 20 Minutes*: Technological Performance, Hedonization, and the Thrill of Impending Failure." In this chapter, the work of Darsha Hewitt is used to show how DIY practice can be performed as both a pedagogical tool and as entertainment for others. Taken together, the chapters that explore unblackboxing in DIY see it as three things: a way to creatively discover the interiors of technology and bend it into something new, a way to make technology more legible and transparent, and finally as a dynamic, performative act.

The next section of chapters (theme 3) looks at DIY practice as a highly personal endeavor that is linked to building one's identity. This third cluster includes chapter 3.1, titled "Identity and *Taratter MI-03*: Device Art, Chindogu, and Alternative Presents," which looks at the work of Maywa Denki as a personalized world-building endeavor. In other words, Maywa Denki uses DIY-style techniques to build out devices from an imaginary parallel universe. Chapter 3.2, "Identity and the *Barbie Liberation Organization*: Culture Jamming, Technical Détournement, and Mediagenics," expands the concept of disobedience through culture jamming. In this chapter, the 1993 work of the Barbie Liberation Organization is explored as an example of how DIY is often deployed as a creative form of protest designed to garner media attention. Next, chapter 3.3 is titled "Identity and the *Stock Market Skirt*: Gender, Telerobotics, and Clothing as Conversation." In this chapter, the Internet-enabled robotic work of Nancy Paterson is used to show how DIY approaches are often used to highlight topics like gender. Taken together, the chapters that explore personal aspects of DIY see it as three things: a path to explore parallel worlds, a way to get media attention, and a way to highlight topics like gender that regularly get excluded from "ordinary" discussions about technology.

The fourth cluster of case study chapters (theme 4) explores the topic of disobedience in DIY culture. Here, do-it-yourself technical production is regularly undertaken as a tactical and political form of protest against different institutions. Chapter 4.1 is titled "Disobedience and *Robot K-456*: Wabi-sabi, Electronic Arte Povera, and Beautiful Mistakes" and examines a 1965 work by Nam June Paik that specifically challenges the idea that robots should be useful servants of humans. "Disobedience and *Hairbrain 2000*: Burlesque Technologies, Highlowness, and Neoretroism" is the title of chapter 4.2, which uses the work of Laura Kikauka to show how DIY practice often functions as a low-fi parody of something refined. In particular, the chapter explores a low-tech parody of virtual reality built in 1993 that is funny, weird, and useful. Chapter 4.3, "Disobedience and *Feral Robotic Dogs*: Hardware Activism, Communities, and Planned Obsolescence," looks at work by Natalie Jeremijenko and her collaborators. Taken together, the chapters that explore disobedient aspects of DIY see it as three things: a way to challenge the idea that technology should solve problems, a way to impede or jam commercial interests, and a method for activists to launch their efforts into the world.

The last and fifth theme addresses the topic of commercial ventures taking the work of noncommercial artists and selling it. This section exists as a single chapter, titled "Theme 5: Selling Out and *Graffitiwriter*: Co-option and DIY Electronics as Unpaid Industrial R&D" highlights the topic of nonprofit DIY projects being appropriated or stolen by industry. It looks at a case of a graffiti-painting robot created by the Institute for Applied Autonomy whose design was taken by Nike and deployed as a part of their "Livestrong" advertising campaign at the 2009 Tour de France.

The book concludes with a summary, recapping the conceptual frameworks and methodologies of DIY technology practices and gives an argument for why we should care about DIY approaches in the first place. A summary of the five themes and their corresponding case studies is presented in "Conclusions: The DIY Mindset" along with a synthesis of the advantages that DIY approaches have in contemporary art and design practices. The core thing to remember is for you to forge ahead and make something: take this book as inspiration and try building something even if it is a hacked mess. It's okay.

In many ways, this research is drawn out of my personal experiences growing up in a rural "DIY" environment, through my hands-on exhibition and fabrication experience as an electronic artist over the last three decades, as a scholar of technology and media culture, and through seeing this style of work reemerge under the label of "making." The aim of this project is to empower DIY electronic practitioners and to aid scholars of art and technology by providing a framework to orient and focus their work. In conclusion, I hope this book brings together the themes of electronic DIY practice, critical theories of technology, and contemporary art to form an innovative and cohesive framework that inspires you to build something. Read and explore, make strange things, and have fun while you're doing it. Take a photo of what you've built and send it my way—I'll actually send you something back as a thank-you, but you'll need to read through to the end of the book to see how that works.

0.1 A History of Electronic Art in the Twentieth Century

This chapter is a story of a full century of "making" before the term emerged, how it evolved, and what trends materialized in the hybrid domain of electronics and art. It tells a narrative of artists as innovative technologists, hackers, and inventors, and in the process starts to sketch out a map for a general history of technology-oriented artwork in the twentieth century. In hindsight, the contemporary maker movement's key mistake was to think of itself as a revolutionary thing, when it was simply the continuation of a fascinating history of creative amateur technologists that reaches back at least a century.

Defining Electronic Art

Within this book, I use "electronic art" as a category that encompasses many different objects and practices. Before exploring specific case studies and themes within creative electronic DIY work, it is imperative to lay a bit of groundwork to explore the history around this field of practice. Part of the purpose of using a somewhat older term—*electronic art* as opposed to something like *new media art* or *digital art*—is to show that this category is still active and is a useful framework to help better understand the social-technological complexity of creative practice. Before information electronics emerged in the middle of the twentieth century, artists had been using simple electrical components since before the 1920s—and the fascinating lineage around this type of work reaches back over a hundred years. In many ways, electricity and artists are good old friends. It's not new media.

Artists gravitate to materials and techniques as soon as they are available—whether paint, sculptural materials, or imaging technologies. There have always been "emerging media" artists who seek out avant-garde materials to expand or redefine their work. As an extension of this experimentation, artists creatively enjoy a century-old history of embracing electricity. Through this history we can understand that their attraction to electrical and electronic media is driven not just by the novelty of technologies, but also by their affordances—interactivity, motion, logic, and aesthetics. New technologies also reflect the changing landscape of popular culture, media, and sociality. As a result, many artists use electronic objects as materials or products to enrich their work by bringing it into contemporary conversations—interdisciplinary conversations that span engineering, design, culture, and other fields.

Objet Trouvé, Ready-Mades, and Picasso

The entrance of electricity into twentieth century contemporary art owes much of its lineage to the interjection of "everyday" found objects into the formal setting of a gallery. The use of everyday objects in contemporary art was initially pioneered by artists like Pablo Picasso, Carlo Carra, and other avant-garde artists of the early twentieth century—or at least this is how it is commonly understood in mainline art history through the narrative of *Objet Trouvé*, literally translated as "found object."[1] Notable pieces include Picasso's *Guitar* (1913), a still life composition of collaged paper, charcoal, ink, and chalk on blue paper mounted on cardboard that includes a front page of a printed newspaper glued onto the surface. This inclusion of day-to-day items marked a significant shift in artistic technique. Up to that point, art had primarily focused on representing exceptional objects on the canvas, rather than literally gluing everyday physical objects onto it. Other artists extended this collage technique to portray more intimate aspects of everyday life by incorporating maps, train tickets, and plastic packaging as materials.[2] The next move—to actually include machines and electronic devices—was propelled by two core ideas: the concept of the ready-made and Dadaism.

Marcel Duchamp coined the term "ready-made" (*Objet Trouvé*) in 1915 "to describe a work of art that consisted of an everyday object, which became a work of art for the simple fact that the artist had selected it."[3] When the term ready-made was introduced into the art world, it was commonly used in the United States to describe manufactured goods. "Ready-made" was synonymous with "off-the-shelf," "factory-built," "ready-to-wear," or "off-the-rack." This is in direct contrast to objects considered artisanal, hand-made, or bespoke—which prior to mass consumer product manufacturing had been more prominent. Ready-mades, in contrast, were intentionally bland, blasé products.[4] While the Futurists of the same era were technologically engrossed in and in wonder of industrialization, Duchamp was already bored with it. He did not think of ready-mades as objects that praised contemporary industrial design and manufacturing, but quite the opposite. The ready-made was part of an effort to battle the aesthetic and "retinal" components of art. He sought to give taste, connoisseurship, and the entire art world the middle finger.[5]

Duchamp's ready-mades were driven by an angry nihilism toward manufactured objects and cultural institutions.[6] This drive was reflected in his first ready-made, the *Bicycle Wheel* sculpture from 1913, made the same year as Picasso's *Guitar* collage. *Bicycle Wheel* was an assembly of a front fork and rim of a bicycle mounted upside-down on top of a painted wooden kitchen stool.[7] Although this particular sculpture was built as a studio prototype and never exhibited, his later ones were installed in galleries. *Bottle Rack* from 1914 was a signed bottle drying rack, and *In Advance of the Broken Arm* from 1915 was a signed snow shovel. These pieces stripped away visual aesthetics and drove home the idea of art as a psychological and social construction. Two years later, Duchamp's *Fountain* (1917)—an inverted and signed ceramic male urinal—earned him notoriety after he controversially put it into an art exhibition.[8] It enraged people because it was drab, industrially produced stuff available at a store, not art.

By claiming that art was whatever an artist chose to be art, the idea of the ready-made gave birth to the entire field of conceptual art and "severed forever the traditional link between the artist's labor and the merit of the work."[9] This move dramatically upended the trajectory of contemporary art in the twentieth century. Beyond bicycle wheels and urinals, Duchamp laid a path for artists to incorporate other everyday materials into their work. After World War I, Duchamp experimented with machines that mechanically produced optical effects. His *Rotary Demisphere (Precision Optics)* machine from 1925 is an early example of electric motors in art. This device features a triangular base that supports tripod-like poles that hold a large plexiglass dome decorated with a nonconcentric circle pattern. A belt is strung between the dome and an electric motor near the base, and when the motor is set in motion, the plexiglass spiral appears to throb and pulsate toward the viewer.[10] Through Duchamp's embracing of ready-mades and found objects, electric motors and kinetics became a more prominent part of the artistic palette.

Dadaism and Good Taste: Shocking Common Sense through Technology

The acceptance of electronics into the formal world of contemporary art was accelerated by the Dada movement during the same period, which had similar antiestablishment motivations to Duchamp. Dadaists were a diverse group of poets, photographers, performers, sculptors, painters, and mashup artists that assembled to protest the violence of World War I in Zurich, Switzerland around 1916. During the First World War, artists, academics, and other left-leaning individuals converged on New York and Zurich to avoid being drafted into battle in their home countries. This concentration had the result of forming anti-war sentiment into distinct subcultures. In particular, many artists opposed nationalism and pride in established cultural institutions, which they saw as factors in precipitating mechanized violence. The First World War saw the deployment of many new military technologies, including grenades, poison gas, artillery, submarines, warplanes, and tanks.[11] When these twentieth-century technologies were combined with a nineteenth-century military strategy and mindset, it resulted in one of the deadliest conflicts in human history.[12] The violence of war upended logic and a belief in the standard way of doing things. As a result, this cluster of artists lost confidence in many of the social and cultural institutions of the time, and saw their work as demolishing the old and beginning again: "We began by shocking common sense, public opinion, education, institutions, museums, good taste, in short, the whole prevailing order."[13]

As the war dragged onward, the Dadaist movement became less of an anti-war protest movement and more of a rally against the conservative conventions of art. In the process, it spread internationally to cities including New York, Berlin, Hanover, Cologne, and Paris.[14] Like Duchamp, Dada artists were not interested in crafting aesthetically pleasing work that fit within the existing structure of art. Instead, they wanted to capsize and drown traditional and complacent artistic conventions. Dadaism's creative anarchy further opened the field of art to everyday objects—including electronics—in order to question the public's nationalistic pride and bourgeois taste. To everyday objects, Dada added the variables of chance, montage, "happenings," assemblage, automatism, and installation—core themes that artists carried forward into an age of electronic media and carried though later art movements like Fluxus. In this way, through people like Duchamp, electricity entered the field of contemporary art as an everyday object, and was amplified by Dada as an antiestablishment force where it continued to challenge the idea of what art is for nearly a century.

From Electric to Electronic: The "End" of the Mechanical Age

"The machine" became a metaphor for modernization during the late eighteenth century and nineteenth century in Britain. It served as a core metaphor for envisioning technological change until the 1960s, when the "electronic age" emerged as a newer metaphor for the contemporary era. *Electrical* devices simply use electricity, while *electronic* devices use electricity to manipulate an information signal. For example, a standard light bulb is electrical, while a radio is both electrical and electronic because it has a considerably more complex ability to manipulate and modify electrical signals as information.

Electric components used in art up until the 1960s were primarily mechanical in nature. Duchamp's *Rotary Demisphere* is a fitting example of more mechanical applications of electricity. The system is mechanical and not informational: the device simply turns on and does not process information. Thinking about electrical systems as machines—in contrast to thinking of work as primarily information-oriented—typified much of the artwork that used such components before the 1960s. Consider Naum Gabo's *Kinetic Construction No. 1* from 1920, which is comprised of a single rod of vertical wire with a balancing weight near its base. The wire is set in motion with a clock spring that produces a slender volumetric shape as it resonates back and forth.[15] Instead of merely evoking motion, like the Cubists or Futurists, this work integrated mechanical movement into art. Similarly, László Moholy-Nagy's *Light Prop for an Electric Stage (Light-Space Modulator)* from 1930 (figure 0.1.1) expands on the concept of kineticism in art. By using an electric motor to rotate several vertically oriented pieces of reflective metal, he produced optic effects when its kinetic surfaces interacted with light.[16] This approach is representative of Kinetic Art, or "a genre of plastic art in which the movement of forms, colors, and planes is used to procure a totality in the process of change."[17]

Artists in the first half of the twentieth century thought of electricity as primarily electromechanical, as distinct from what we now call electronic. This can also be called *electric-mechanical* versus the newer

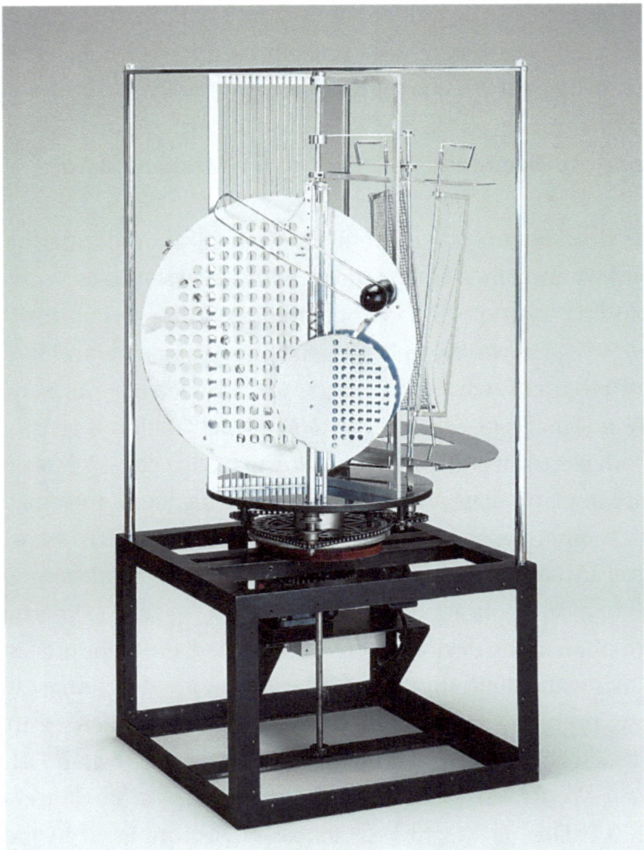

Figure 0.1.1
Kinetic mechanical movement of sculpture emerged as a key feature of art using technology that predated the emergence of information technologies. László Moholy-Nagy, *Light Prop for an Electric Stage (Light-Space Modulator)*, 1930. Photo © Hattula Moholy-Nagy / DACS 2007.

electronic-informational. I define an *electrical* device as one that simply turns on and off, such as a light bulb or an electric motor. The device may be able to be dimmed or slowed down, but it has no feedback or logic to regulate its own control. Electrical devices simply use electricity as a power source. In contrast, *electronic* devices are capable of changing behavior on their own or are able to represent and manipulate information. Electrical circuits have no decision-making capability, but electronic devices do. In this sense, televisions are electronic, while electric motors are not.

Cybernetics, Communication, and the Birth of the Information Age

The advent of cybernetics was a key paradigm shift that ushered in the idea of the information age. Cybernetics, as defined in 1948 by Norbert Wiener, is "the scientific study of control and communication in the animal and the machine."[18] This idea emerged out of communication, which was concerned with how information could be encoded, transmitted, received, and decoded in systems like telephone networks.[19] Wiener's breakthrough was that cybernetics envisioned information as an independent thing—in other words, a signal that was independent of the transport medium. As a result, information became understood as a malleable and transferable entity. Art historian Edward Shanken astutely notes: "By disembodying information from its material substrates and reducing it to a generic signal (as in analog waveforms or the binary code of digital computers), information theory offered a flexible and practical method that could be applied to immediate engineering concerns . . . Cybernetics used it to theorize parallels between the exchange of signals in electromechanical systems and in the neural networks of humans and other animals."[20] Information became a thing that could be electrically transmitted, manipulated, encoded, and decoded through different devices.

Cybernetics also marked a shift in the understanding of how information works on a network level. Information could now be viewed as circulating and dynamic, able to be sculpted, governed, and fed back to control system behavior—what communication scholars would call a "feedback loop." The etymology of the word "cybernetics" points to this. The term was coined from the Greek word for "steersman," or *kubernetes*—the same root of the English word "governor."[21] This ability to "govern" information is a fundamental difference between the electric and the electronic. On its own, an electric device cannot process information or data—only an electronic device can. With this in mind, the electronic age is contiguous with the information age. In 1982, for example, Walter Ong referred to the electronic age as including the media technologies of radio, television, telephones, and satellites that harkened, "the electronic processing of the word and of thought."[22]

The start of the electronic-informational age was also reflected throughout the 1960s in the mainstream art community. By the end of the decade, The Museum of Modern Art (MoMA) claimed that the mechanical age was drawing to a close. On November 27, 1968, a major exhibition titled "The Machine as Seen at the End of the Mechanical Age" was launched.[23] This exhibition looked back on the machine in art through a sprawling collection of artists from Leonardo da Vinci to the contemporary video artist Nam June Paik.[24] Swedish curator Pontus Hultén saw this retrospective as a commentary on how artists of the Western world viewed technology, and as an exploration of how humans have regarded machines as alternately utopian or dystopian.[25] The title, however, indicated that this era of machines was at its tail end. This desire to identify the new information-oriented epoch spanned museums internationally. At virtually the same time, the Institute of Contemporary Arts (ICA) in London launched an exhibition titled "Cybernetic Serendipity" that was curated by Jasia Reichardt. Instead of just alluding to an electronic future, this exhibition fully embraced it. The show featured algorithms and devices for generating music by Peter Zinovieff, artistic computer projects by Gustav Metzger, computers that could generate essays and poetry, and radio-controlled robots by Bruce Lacey and Nam June Paik. "Cybernetic Serendipity" also featured computer-generated movies and computer graphics by John Whitney, and interactive sculpture projects with complex sensors and feedback systems.[26] Electronic and informational media were rising as the new paradigm.[27]

It is also worth emphasizing that the rise of information technologies did not result in the disappearance of electromechanical systems: electromechanical systems simply worked alongside information systems, and artists continued to work with both electrical and electronic materials. The considerable amount of excitement around the processing of information through computers and interactive systems that incorporated feedback is historically monumental, however.

Art exhibitions in the 1960s were also driven by the crucial role that electronics and information systems had in popular culture. Stewart Brand, founder of the *Whole Earth Catalog,* wrote that "the counterculture saw in cybernetics a vision of the world built not around vertical hierarchies and top-down flows of power, but around looping circuits of energy and information."[28] This widespread interest was precipitated by advances in consumer electronics and the rise of cybernetic theory in the 1950s that envisioned information as a transferable and malleable object. Consider Jimi Hendrix's iconic, squealing "The Star-Spangled Banner," where electric guitar feedback took the role of a counterculture anthem at Woodstock. The year 1968 was a pinnacle for protest worldwide, and these moments of political upheaval also influenced the artistic understanding of electronics in art. As a result, the information age erupted into the art world during the second half of the 1960s and continued in the subsequent decades.

Electronic-Information Overload: Too Technical for Aesthetics, Too Artistic for Science

Electronics were adopted by Gary Hill, Doris Totten Chase, Shigeko Kubota, Bill Viola, and other artists using video. Artists also continued to use simple electrical (not electronic) objects in their work, like Dan Flavin making minimalist installations using off-the-shelf fluorescent lights (figure 0.1.2).[29] However, many in the art world were not swept up in this shift toward embracing electrical and electronic media. Although it launched several electronic artists into the mainstream art world, most artists that primarily used electrical and electronic technologies remained within their own subcultures of festivals, galleries, curators, and publications. Although the ICA and MoMA shows of 1968 were significant, they were more of an indication of a shift in popular culture toward consumer electronics and industrial and academic research in information systems—they did not mark a wholesale shift in the art world toward embracing electronic media.

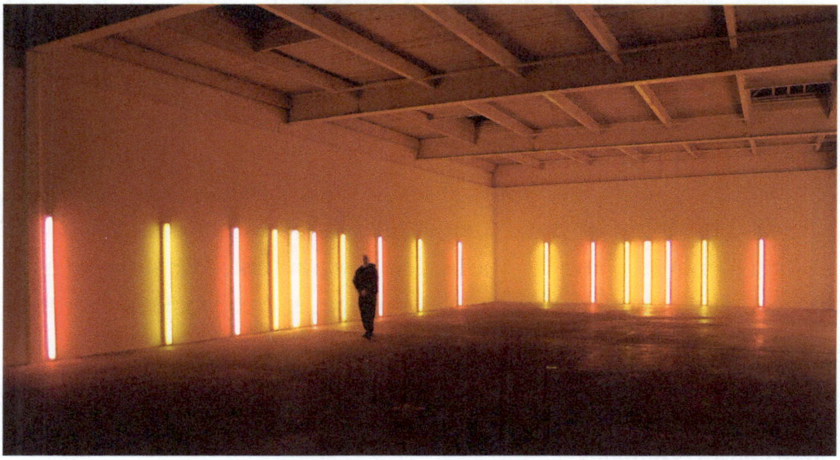

Figure 0.1.2
Dan Flavin regularly built minimalist light installations using off-the-shelf fluorescent components. Dan Flavin, *Alternating Pink and Gold*, 1967. Photo © Estate of Dan Flavin / SOCAN (2021).

Art historian Edward Shanken, for example, believes that electronic art was "too technological to be appreciated under conventional canons of aesthetics, and too artistic to be appreciated according to the criteria of science or engineering."[30] It essentially remained a small-scale, do-it-yourself endeavor that operated on the edges of many disciplines, including art, commercial technological development, and hobby electronic culture, with an occasional intersection with experimental computer science or robotics.

There are also pragmatic curatorial reasons behind the history of electronic art being left out of the art canon. Experimental interactive systems and custom-built electronic projects pose many problems for museums and galleries that primarily exhibit and maintain static paintings and noninteracitve sculptures. Electronic-oriented art projects frequently break down and require maintenance due to their cutting-edge and underfunded DIY development processes. "Cybernetic Serendipity," for example, racked up $40,000 in repairs and technical maintenance between shows, which prevented it from being exhibited at the Smithsonian Institution.[31] As a result, the electric and electronic media that permeated the mainstream exhibition system were primarily noninteractive and fit well within the traditional infrastructure of white-walled galleries. This small number of electronic art successes were easy to install, maintain, and repair. Jon Ippolito argues that an artist like Dan Flavin was able to navigate and succeed in the mainstream art world because he used simple electrical components that were inexpensive and easy to install (figure 0.1.2). In Flavin's case, he was proud that his ethereal lightscapes were built out of low-priced fluorescent tubes available at any well-stocked hardware store.[32] Electronic devices came into contemporary art in the 1920s by way of the ready-made item, and this is the level of consumer-grade commitment that most art institutions were willing to provide. However, the bespoke electronic projects spawned many new opportunities for creation, participation, and collaboration for those willing to bend or build institutions.

E.A.T. in the 1960s and 1970s: The Experimental vs. the Ready-Made in Art and Technology

One approach to addressing the technical difficulty of building custom electronic devices was to connect artists with industrial, mechanical, and electrical engineers. The group Experiments in Art and Technology (E.A.T.)

was formed in 1966 to serve as a matchmaking service between these two universes. The initiative was driven by Billy Klüver, who worked as an engineer at Bell Labs on crossed-field backward-wave magnetron amplifiers. He had already provided technical support to a number of high-profile contemporary artists in New York City, including Andy Warhol, John Cage, Jean Tinguely, Robert Rauschenberg, and Jasper Johns.[33] Klüver had helped them make self-destructing machines, custom-built wireless audio units for sculptures, neon sign paintings, and floating sculptures. E.A.T. was set up to facilitate more artist/engineer partnerships in this spirit.[34]

Experiments in Art and Technology grew out of a collaborative artist/engineer event titled *9 Evenings: Theatre and Engineering* that Klüver and Rauschenberg organized during October 1966 at the 69th Regiment Armory in Manhattan. The project consisted of ten artists and approximately thirty engineers that worked over the course of ten months to develop a mixture of new technologies with avant-garde theater and performance. The event was eviscerated in the popular press with headlines like "Audience Endures a Depressing Spectacle" and statements like "It was such a sad failure, such a limp disaster. . . . It is merely depressing to see such pioneering being done by what appear to be bumbling amateurs."[35] Although the individuals were arguably "expert" artists and engineers, the drastically interdisciplinary collaborations were thoroughly "DIY" in spirit: projects often struggled to function at all, and the ruggedness of the systems and the mode of the audience interaction was not thoroughly developed. Not dissuaded, Klüver and Rauschenberg formed E.A.T. about a month later with the help of Bell engineer Fred Waldhauer and artist Robert Whitman.[36] The organization rapidly expanded in size, and by 1970, E.A.T. had twenty-eight regional chapters throughout the US with approximately 5,000 members, half of whom were artists and half of whom were engineers.[37]

At its core, E.A.T. believed that artist/engineer collaborations would help society by humanizing emerging technologies. This echoes Marshall McLuhan's prompt for artists to act as "antennae" to evaluate new technologies.[38] Artists could add "individual variety, pleasure and avenues for exploration and involvement in contemporary life" through new technology in collaboration with engineers.[39] This humanization was not primarily aesthetic but oriented around the idea that the "artist expands our vision of the world and inspires the content of the communication media as well as the form of the media themselves."[40] As a byproduct, it was also proposed that these collaborations could help industrial partners generate future-oriented thinking within their companies and ameliorate fears the public had over rapid technological change. Accordingly, E.A.T. pitched itself as a cultural middle ground between the progressive countercultural left and industry, as "a mutual agreement in order to avoid the waste of a cultural revolution."[41] However, even for projects that had funding to make the work more technically robust—like E.A.T.'s development of the Pepsi Pavilion at the 1970 Osaka World Exposition—the close ties to large multinational corporations led to a clash of ideals.[42] The relationship between Pepsi's management and E.A.T. disintegrated after a disagreement over having a structured schedule of events or an open-ended and experimental series of happenings in the space.[43]

Electronic art, for the most part, is a theater of do-it-yourself practitioners experimenting with technically unstable devices. Although E.A.T.'s intentions were organized and ambitiously professional, the delivery of their projects was widely derided by art reviewers as being a mess: "If the American engineers and technologists participating in this performance were typical of their profession, the Russians are sure to be first on the moon."[44] Critics felt that the work fell technically and artistically short. The E.A.T. organization thus illustrates a larger trend that has challenged the popularization of electronic art. Projects that are defined as small-scale and experimental tend to succeed in subcultural settings, while larger-scale projects are reviewed positively only when they meet the same technical stability requirements as consumer electronics. As a result, much of this work has thrived as a subcultural DIY practice.

Antenna: Artists as Sociotechnical Researchers

The work of Marshall McLuhan during the 1960s was also influential in shifting mindsets from the electric and mechanical to the electronic and informational. McLuhan's written work in media and communication centers on the rising tide of the electronic in Western civilization over the decade. In his popular books, he argues that this flood of electronic communication is radically reconfiguring society. In *The Gutenberg Galaxy*, for example, McLuhan suggests that the visually and individually oriented culture of print is in the process of being replaced by electronic media. As a result, a more collectively oriented "global village" has emerged, focused on sound and speech.[45] *Understanding Media* further outlines the immense impact that electronic systems have on culture. McLuhan goes so far as to proclaim that they upend the existing global order: "After more than a century of electric technology, we have extended our central nervous system itself in a global embrace, abolishing both space and time as far as our planet is concerned."[46]

McLuhan's influence impacted several practitioners in the arts, especially the field of artists working with new technologies. McLuhan specifically steered artists away from self-expression and toward exploring new technologies in an effort to predict social problems. In the introduction of his second edition of *Understanding Media* published in 1964, he put forward the argument as follows: "Ezra Pound called the artist "the antennae of the race." Art as radar acts as an early warning system, as it were, enabling us to discover social and psychic targets in lots of time to prepare to cope with them. This concept of the arts as prophetic contrasts with the popular idea of them as mere self-expression. . . . We can afford to use portions of them [the arts] that enhance the perception of our technologies, and their psychic and social consequences."[47] In turn, several artists adopted this exploratory, antenna-like approach to researching new communication technologies regardless of their professional technical experience. In addition to experimental organizations like E.A.T. that encouraged radical interdisciplinary collaborations, McLuhan's approach of artists as technical researchers inspired and fueled many organizations that emerged in the 1980s, like Ars Electronica, V2_, Transmediale, Interaccess, ZKM, and others.

The 1980s: An Institutionalization of the Technologically Tentative

As time passed, the genre of electronic art—lacking formal institutional representation—congealed into international subcommunities. The period of the 1980s saw the experimentation of the 1960s and 1970s congeal into a number of relatively stable local organizations that encouraged technologically oriented practices. In Austria, for example, the yearly Ars Electronica Symposium exhibition and conference began in Linz in 1979. During the same period, other similar organizations began to emerge to foster artists that wanted to significantly engage with information and media technologies. Other initiatives include the V2_ Organization / Institute for the Unstable Media. It started in 1981 in the Dutch town of 's-Hertogenbosch as an interdisciplinary artists' squat. Out of frustration at the visual arts community, in 1987 it began to focus exclusively on electronic art and other "unstable media."[48] Founders Alex Adriaansens and Joke Brouwer saw the instability of electronic art as an advantage. Happenings and installations became their core strength and brought these artists more in touch with playful and interactive mass culture (figure 0.1.3).

In Canada, InterAccess in Toronto grew out of the Artculture Resource Centre (ARC), founded in 1981. The organization brought together artists that were using Telidon, a Canadian pre-web interactive telecommunications system somewhat similar to videotex or Minitel. Initially launched with a solo show by Brian Eno in 1981, the organization grew to include a core group of artists including Nancy Paterson, Tom Leonhardt, Jeff Mann, Graham Smith, Nell Tenhaaf, Paul Petro, Doug Back, and David Rokeby.[49] Other institutions founded during this period include the Transmediale festival in 1988 in Berlin and the Zentrum fur Kunst und Technologie (ZKM, "Centre for Art and Technology") in 1989 in Karlsruhe, Germany.

These organizations were driven by the frustration electronic artists had with being neglected by mainstream visual art during the 1970s and 1980s. Alex Galloway notes that the contemporary art journal *October*, which can be viewed as a yardstick for artistic and intellectual trends, did not even cover the field of art and technology until 1985.[50] And, as previously mentioned, museums were reluctant to embrace projects that incorporated electronic media. As a result, electronic artists often labored in silent obscurity. According to Edward Shanken, "even Paik, the most celebrated artist associated with Art and Technology, struggled well into the 1980s."[51] Organizations such as Ars Electronica, V2_, and InterAccess supported artists, helping them to continue making groundbreaking interactive and intentionally unstable work that bucked institutional demands. These organizations often worked as collection points for DIY practitioners—a bit like proto-hackerspaces—that aided in the production and distribution of this work.

Electronic art started to ride on separate circuits through distinct exhibition styles, approaches, and events. Many artists that used electronic technologies viewed their work as part of a dynamic process, rather than part of the "serious and stable" institutional model. This process-based view often clashed with galleries interested in the preservation and presentation of static objects. Instead of maintaining a hermetic white gallery, electronic art organizations created spaces where projects were designed to be touched, make

(a)

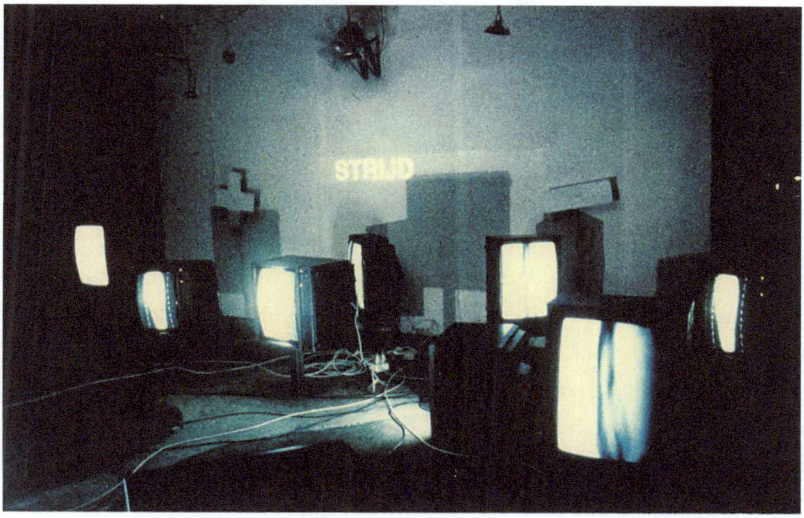

(b)

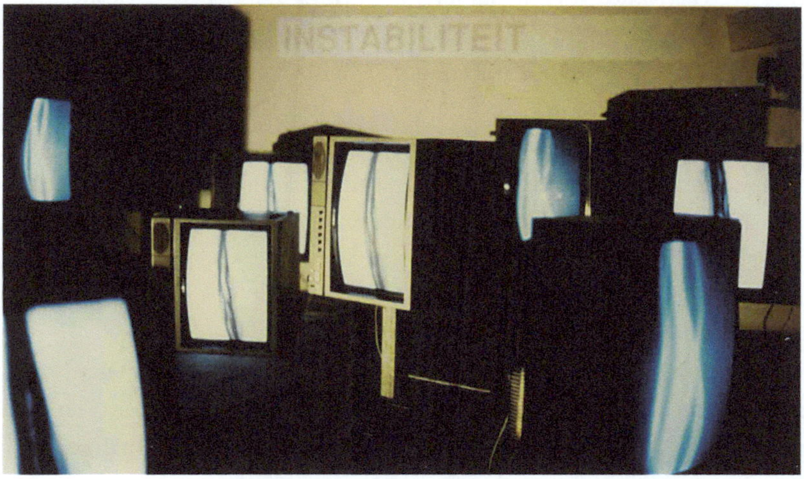

Figure 0.1.3

An example of one of the numerous projects at V2, which operated as an organization, exhibition and performance space for "unstable media" in the Netherlands since 1981. Joke Brouwer and Alex Adriaansens, Installation for the Unstable Media, 1988. It consisted of a number of interactive monitors in two different rooms. In one space at V2_, the audience charged a virtual capacitor, which was discharged upon entry into the second space. This change impacted the display of the sideways-positioned CRT screens scattered through the environment. The title gives reference to V2_'s new vision (circa 1987) of focusing on electronic art and other forms of "unstable" media. Courtesy Jan Sprij.

noise, and break down.[52] At times, these electronic art performances more closely resembled a music venue than a gallery. These performative and interactive spaces required electrical infrastructure and staff to help start up projects, keep them running, and shut them down. With an emphasis on interactivity, these initiatives often drew historical inspiration from avant-garde and interventionist art movements like Fluxus and Dada along with technology-minded theorists like Marshall McLuhan. To this day, numerous art festivals focus exclusively on electronic art, like Ars Electronica, the International Symposium of Electronic Art (ISEA), the Dutch Electronic Art Festival (DEAF), and the *Festival Internacional de Linguagem Eletrônica* (FILE) in Brazil. In response, some mainstream institutions invested in substantial support for art and technology initiatives in a museum setting. For example, the Centre Pompidou in Paris integrated engineers as part of their maintenance and preservation efforts by the late 1990s.[53] Inclusion of DIY electronic work in larger institutions happened occasionally, but the work primarily thrived in small, subcultural communities that regularly blended art with engineering, design, science, and information technologies. Through a slow grassroots evolution over the course of the twentieth century, electronic art has developed a distinct cultural approach to doing and appreciating art; most of it is best understood as a *variegated* DIY process and *mindset* instead of a stable and institutionalized canon.

Electronic / Media / Digital / New Media

Many seemingly interchangeable terms are used to describe artworks that incorporate technology: "electronic art," "media art," "digital art," and "new media art." Although there is considerable overlap, it is worthwhile to pause and consider what the specific term "electronic art" captures in comparison to seemingly similar concepts. As I have argued above, electronic art is the oldest of this cluster of terms, and it encompasses all work that primarily uses electronics as the medium of studio production.[54] Media art, on the other hand, is defined as artwork that uses communication technologies like video. The Canada Council for the Arts, for example, refers to media art as artwork that depends on a technological component to function. Harkening back to the informational turn of the 1960s, they particularly consider new media and moving image technologies used to transmit and store information to be media art.[55] In art and technology circles, the "media" of media art is often viewed in a McLuhanesque sense as any new technology that extends perception.[56] In more mainstream galleries and institutions, "media" refers to a general understanding of forms of mass communication. By comparison, digital art refers to any work that relies on computation, and new media highlights work after approximately 1990 that is computer-oriented, interactive, or Internet-based.[57] Sarah Cook and Beryl Graham define new media art as "art that is made using electronic media technology and that displays . . . interactivity, connectivity and computability, in any combination."[58]

These flexible labels necessarily build on each other, overlap, and are hybridized. Cook and Graham's definition of new media art highlights how these terms are constructed from earlier iterations. From a material perspective, new media is made out of digital technologies, the digital is built with media technologies, and media technologies are built from electronics. However, in social contexts and in wider culture, labels and categorizations shift, depending on who is seeing the work. In other words, the same piece of DIY technology can be both a piece of electronic art and a work of critical design; it often simply depends on the audience. In large part, this branding is more the domain of curators, writers, and gallerists than artists. Often, the coining of a new term—like "Holopoetry," "Biotelematic Art," or "Transgenic Art"—is used to lay claim to a niche subdomain. The person that invents the term often advantageously positions themselves as the first pioneer of the new genre.[59] As a result, these micro-labels are of little general use because they only include a few people across the planet. Although the term "electronic art" could apply to any project incorporating electronics, I use it to refer more specifically to individuals that use electronics as a medium or material, and use their art to question technology's role in culture. As a result, I argue that electronic art is still useful as a category—especially with the increased interest in creative electronics that has emerged since 2005 under the banner of maker culture.

The "Effect" Effect of New Media

In many senses, the term "new media" is worn out and is no longer of much practical use. Beyond this, the term can be counterproductive in highlighting historical trends in art and technology that span back at least a century.

A focus on the term "new media" skews our perception of this work, and reverting back to the older term "electronic art" maintains more continuity in the field of practice. This is not to say that newer fields—like computation in the arts, algorithms, blockchain, or networks—are not of consequence in contemporary art practice. My point is more to emphasize the continuity of electricity in art as a significant field of practice for about a century. The medium of electronics is *both old and new*. However, continuing to refer to this work as "new" media fails to consider its longer history.

The term "new media" was deployed in the 1990s to differentiate electronic and digital work from noninteractive and nonnetworked pieces, both in traditional studio art and in noninteractive works using electrical components. Although the term "the new media" was used as early as 1966 to describe computer-based media within the framework of educational technologies (Rossi & Biddle), by the time the World Wide Web became popularized through the release of the Mosaic browser in 1993 and Netscape Navigator in 1994, the term had dropped the "the" and was simply "new media."[60] With the ascent of the Internet in the mid-1990s, the field of electronic art largely embraced *new media* as a term and a mindset. Digital newness was the best way to differentiate electronic media from older forms of artistic expression. This orientation toward newness resulted in what Rafael Lozano-Hemmer termed "the 'effect' effect": technological novelty and special effects became the core object of the artwork and its primary incentive.[61] Artwork using new technology is often an exploration of the new effects available within the medium.

Exploring the potential of new technologies is not new. In some ways, it is a well-worn and significant tradition within art practice. Special visual effects have been implemented in art for centuries through Mannerist anamorphism, Baroque ornateness, Modern collage, and Op Art. The criticism that Lozano-Hemmer brought forward to new media art in the 1990s was that this work is often arrogantly superficial and lacks reflective distance from the technology it is using. It perpetuates the myths that new technology is "(1) providing a truly global culture; (2) introducing infinite creative possibilities; (3) to be trusted with the management of resources both 'natural' and 'human'; and (4) eradicating discrimination on the basis of gender, race, or class."[62] In other words, new technology is often portrayed as a cure-all for human problems, and during the early 1990s when information technologies, computer networking, and virtual reality were rapidly expanding, these technologies were often sold with the promise of being able to radically expand human creativity and to bring about social equality. This was partially true, but was a drastic oversimplification and a "technosolutionist" approach to complex human problems.[63] Lorne Falk referred to this in conversation with Lozano-Hemmer as being "technologically correct," alluding to the pejorative term "politically correct" that the Right coined to describe the Left's attempts to be inoffensive in America since the 1980s.[64] The technologically correct, or technosolutionist approach, generally believes that more technology is better for society—that technology is the answer to human problems—and is often a sales technique promoted by commercial technology companies that is still prevalent today.

Exploring the expressive potential of new technology is not negative on its own. In fact, such experimentation crucially uncovers how emerging technologies impact how we perceive the world, communicate with each other, and shape our identities. For example, E.A.T.'s mission in 1966 was described as helping to "guide the artist in achieving new art through new technology." The problem arises when artists and curators uncritically buy into the emancipatory narrative of technology; their work lacks a reflective depth. There is often little difference between new media artwork and a commercial technological demo where we are invited to admire a computer system's improved capabilities, resolution, or speed. Worse, new media can reinscribe existing divisions in who gets to benefit from technology. It can resemble "toys for boys"—reviving the worn-out colonial archetype of the masculine pioneer. The first (usually male) person to smash through a new technological boundary is rewarded, irrespective of whether their work has anything culturally interesting to say.[65] Sarah Diamond sees this style of work as providing "an ecstasy of technical effects, formal functionality, and coolness, without social context."[66] In my mind, this is part of what new media art represents.

An overemphasis on the newness of media creates lackluster art projects that are also overwhelmingly mediocre from an engineering standpoint. Artists have horrible difficulty making technologies at the level of industrial engineers and commercial industry. As a result, they need to either employ a DIY approach to build new tools or use consumer-grade technologies that are available to the general public. The former is often technologically frail, and the latter is more an innovation of use than a radical technical development. Several artists are equally talented at engineering;

David Rokeby, Char Davies, Seiko Mikami, Natalie Jeremijenko, Krzysztof Wodiczko and Myron Krueger were innovative artist-engineers active in the 1990s. However, as historian Edward Shanken states, independent artists are generally "unable to compete on a technological basis with the spectacularity of scientific demonstrations, mass media, or Hollywood special-effects."[67] New media art that falls into the "effect" effect often sells us on novelty and neglects McLuhan's prompt to help society understand and rethink its complex relationship to technology.

The Web 2.0 Pivot: The Art Formerly Known as New Media

If a 2005 exhibition at Walter Phillips Gallery at the Banff Centre was any indication, the post-Y2K period saw a reevaluation of new media. "The Art Formerly Known As New Media," cocurated by Steve Dietz and Sara Cook, provided a retrospective of electronic artwork on the tenth anniversary of the Banff New Media Institute. The exhibition marked a clear shift in electronic art and new media elsewhere. In particular, 2003 saw the Walker Art Center, the Guggenheim, and SFMOMA scale back their new media initiatives. In a wider social context, the Internet had started to shift into its "Web 2.0" phase of user-generated content and social networking: for example Wikipedia was released in 2001, MySpace came in 2003, Vimeo and Facebook arrived in 2004, and YouTube launched in 2005.[68]

It should not have come as a surprise that the early 2000s were when digital new media began to lose its status as an avant-garde form. In 1999, critic and educator Peter Lunenfeld proposed that the whole rubric of new media should only be temporary. He argued that "the very term 'new media' is ambiguous. . . . In the end, the phrase 'new media' turns out to be yet another placeholder, this time for whatever we eventually agree to name these cultural productions."[69] New media is a temporal term, now dated and reminiscent of the Y2K era.

Critical Design: The "Effect" Effect is the Enemy

Emphasizing the novelty of new media has done significant damage to art practices that strive to comment on the role of technology in everyday life. Critical designer Anthony Dunne, for example, dismisses the entire category of electronic art as a useful place for reflection about the impact of technology on everyday life. He sees the field as driven by glamorization, "concerned only with the aesthetic expression of technology for its own sake."[70] This is what Lozano-Hemmer refers to as the "effect" effect. I agree that criticizing the glamorization of technology is worthwhile. After all, Dunne and Lozano-Hemmer are, by proxy, expressing wariness about easy promises of utopianism and the progressiveness intertwined with concepts like new media. However, to describe all of electronic art as being concerned only with technological aesthetics is blatantly inaccurate.[71]

Art is a porous category. The history of electronic art that I have outlined shows how electronic objects can float between communities in art, design, computer science, the humanities, and industry. For this reason, grassroots organizations of electronic artists vigorously carved out their own subcultural terrain in the 1980s and 1990s. As a result, I see that electronic projects themselves have a form of agency and can act as a meeting point between disciplines.[72] Here, Dunne's term "electronic objects" is useful in order to draw attention to individual projects instead of disciplinary distinctions. He writes that "the electronic object is a confusion of conceptual models, symbolic logic, algorithms, software, electrons, and matter." However, it is false to think that these designed objects cannot also function as art. The inverse is also true: many artists also clearly do engineering.

The remainder of this book takes Dunne's dismissal of the category of electronic art as a challenge. The field is stacked with groups and projects that question the role technology plays in contemporary society. Accordingly, I work to reframe electronic art by subtracting technological novelty from the narrative. When the progressiveness and optimism of new media is subtracted from the narrative, a different form of electronic art emerges—one that is critically engaged with technology, technically uncertain, and often a do-it-yourself endeavor.

Conclusions: Electronic Art as DIY Practice

Electronic art can be thought of as encompassing and including any piece of art that has both electricity and information flowing through it, including video art, work with computers, custom electronics, or electronic music. Despite being a historical term, the term is still useful in highlighting work that focuses on the medium of discrete electronics and "device art," especially within the contexts of the rise of the Arduino and consumer 3D printers, and the rejuvenation of electronic prototyping through maker culture since 2005.[73] Having had at least a half century of lead time ahead of the concept of the maker movement, electronic art provides a rich and varied field of creative practice in making and electronics.

In summary, electricity has been part of the sphere of art by way of artists incorporating ready-made consumer items into their work since at least the 1920s. A cybernetic shift from the electric and mechanical to the electronic and informational occurred in the mid-1960s, and the formation of organizations solely focused on art and technology took hold from the 1960s through the 1990s. The term "new media" was mobilized in the 1990s, but this term—and the novelty of interactivity and the Internet—began to be widely adopted by mainstream institutions around 2005. At the same time, 2005 gave birth to the archetype of the "maker," which integrated and popularized some components of DIY electronic culture. However, the maker movement provides us with few details about history, sparse details on why we should make something beyond technical curiosity, or where we should go in the future.

Art and design, on the other hand, are relatively obsessed with their own histories and futures, and provide a plethora of ideas, concepts, and terms that point a clear path forward when combined with scholarship on technological development. In hindsight, it is useful to historicize and rethink the work done in electric and electronic art over the past century through the lens of amateur-developed and "punk" technologies: it provides many useful signposts for contemporary maker culture and gives important clues about the unexpected and surprising advantages of technological design from artistic and nonindustrial perspectives.

0.2 A Definition of DIY: Do + It + Yourself

"DIY" is ambiguously undefined as a category despite being a widespread, ordinary, everyday occurrence. According to the New Oxford American Dictionary, DIY, or D.I.Y., is simply an abbreviation for "do-it-yourself." On its own, this definition provides little clarity. Some theorists like Florian Cramer have asked whether the term DIY actually means anything at all, and they suggest that it is "best understood from within, since it includes personal involvement and entanglement."[1] Other theorists, like Julia Lupton, talk about DIY graphic design as being able to be effective "not only to be *part of* a public, but also to *have* a public, to address an audience through acts of deliberate, designed, expression."[2] Although a number of individuals have worked to define the term of DIY, it largely remains an untheorized activity done by millions of individuals on a daily basis, and not a well-worn domain of academic scholarship. As a result, carving out a general definition of DIY beyond the scope of art, design or technology is useful to bring clarity to a diversity of everyday cultural practices.

In North America, "DIY" often brings to mind home improvement stores like The Home Depot (figure 0.2.1) or Lowes that assist individuals in repairing or upgrading their residences. In Europe, DIY is also synonymous with home improvement retailers, with Statistica listing the leading "DIY retailers" of 2019 as the following home improvement companies: Groupe Adeo (France, with €21.3 bn of revenue), Kingfisher (UK, €13.1 bn), and Obi (Germany, €6.8 bn).[3]

However, it also often refers to fabric shops, scrapbook supply stores, hobby stores, and car part suppliers. Although locations like these attract many professionals, the appeal of DIY is that a job can be done cheaper and in one's own time frame. Notable DIY electronic retail outlets from

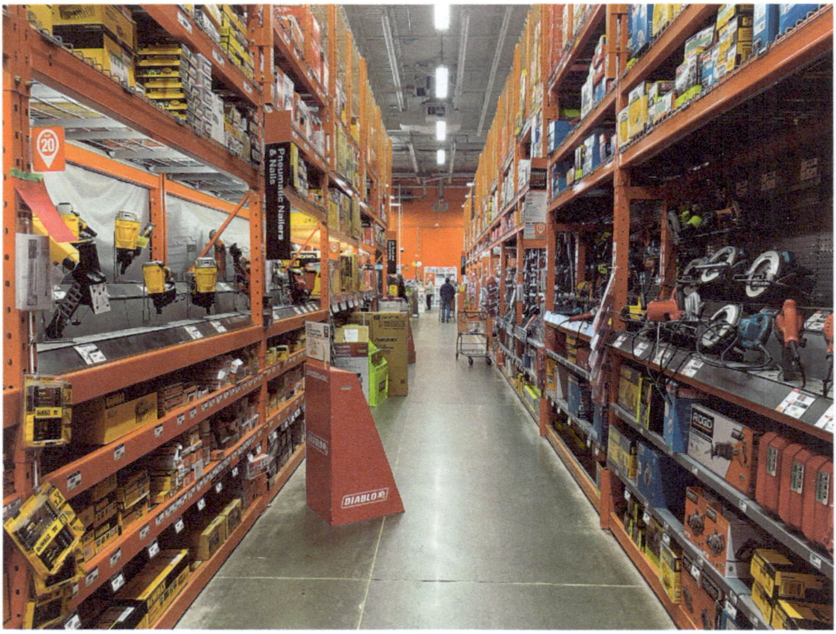

Figure 0.2.1
In popular culture, DIY often refers to stores like Home Depot. Courtesy Kaiwen Yang.

history included Radio Shack, which sold electronic components for tinkerers to modify or create their own electronic products. In general the common assumption is that "doing it yourself" makes the person using the object responsible for making, repairing, or modifying it. However, DIY is significantly more than this—and it has many interesting historical, psychological, and practice-based implications that are worth unfolding. DIY is technically and culturally much, much more than self-repair without the use of experts. Let me explain.

As a foundation, several cultural and artistic movements can be thought of as being DIY oriented. These include the historical punk movement, Fluxus, mail art, and even phone phreaking—the act of using custom electronic devices to get free long-distance phone calls. Punk embraced a bricolage of hacked fashion and a learn-while-you-play approach to music, while Fluxus often explored art as an ephemeral happening or set of instructions that anyone could do. Mail artists experimented in a similar fashion with things that could be sent through the postal service, which placed more emphasis on a peer-to-peer network of individuals instead of a centralized institution like a gallery or a museum.[4] These threads and the interrelationships between different forms of DIY practice will be explained in the pages and chapters ahead.

In this chapter, I break apart the constituent terms of *Doing*, *It*, and *Yourself* to build a definition of DIY. From there, we can consider what it means for artwork to be considered "DIY." Each subsequent chapter of this book continues to add more lexicon around this core definition of DIY—after learning the core of what DIY is in this chapter, the functional attributes and mindset of DIY culture is articulated through in-depth examples—but let's start with the basics of what we mean when we say "DIY," or "do it yourself."

Defining DIY: Contexts of "Do"

"DIY" starts with a "D" for "Doing"—and unpacking the word helps provide a richer definition of the concept of DIY. To start, doing emphasizes action over planning: it is *do* it yourself, not *plan* it yourself. Because of this emphasis on action, the DIY process embraces jumping into an activity without hesitating to think about formal qualifications or past experience. DIY inherently involves actively manipulating tangible objects. For this reason, I base my concept of DIY on Paul Dourish's notion of action as involving "the creation, manipulation, and sharing of meaning through engaged interaction with artifacts."[5] DIY, in other words, is an engaged interaction with physical artifacts. To this basic understanding, I add that DIY involves a particular *kind* of action, one that is intrinsically rewarding, embodied, and process oriented.

Doing Is Intrinsically Rewarding

Generally speaking, DIY practice believes that there is value in manual labor. As philosopher Michael Crawford argues, the manual labor wrapped up in doing has a strong cognitive, social, and psychological appeal.[6] Similarly, sociologist Richard Sennett illustrates that manual labor and the pride that comes from "doing" have been sidelined by contemporary capitalism. Both Crawford and Sennett emphasize craftsmanship, which is defined as doing something well for its own sake.[7] Craftsmanship and DIY both entail manual labor done by an individual, but they drastically diverge in terms of the finished quality of the work. The doing of craftsmanship is built upon skill, mastery, and artisanal expertise developed over time. "Crafting is about a way of being in the world that requires not just knowledge but practice"[8] The doing of DIY practice is the opposite: it requires no technical skill or experience. In this sense, DIY could be described as nonvirtuosic, more about being in the world by attempting to create things. However, I argue that DIY undergirds craftsmanship, rather than being an inferior alternative. In fact, I see DIY as both a precursor and subset of skilled craft. DIY involves the beginning phase of trying something new because it is fulfilling, which may develop into craftsmanship over time as a result of experience.

Doing Is Embodied

"Doing," in both DIY practice and craftsmanship, is an embodied activity where hands and mind work together. Although DIY can be used to describe learning computer programming or other mental skills, DIY "doing" generally places emphasis on work done physically. In the process, it de-emphasizes the traditional Cartesian split between mind and body. This dualism sees reality as consisting of the two independent components of mind and matter. Cartesianism is echoed by early cyberpunk science fiction authors like William Gibson, who describe reality as taking place in cyberspace, or the "nonspace of the mind"[9]. This dualism prioritizes the mind and intelligence, while the flesh of the body is seen as a limiting force and referred to as derogatory "meat."[10] DIY "doing" is more body oriented, with action taking lead over abstract thought. It embraces "meatspace," or real-world physicalness, through action—with hands leading the way. It is the engagement of an individual as a physical being directly working on a particular task. In this way, DIY shares common ground with research in computer science like ubiquitous and physical computing that downplays screen-centric "nonspace of the mind" platforms like virtual reality.

Doing Is Process Oriented

DIY practitioners regularly take pride in the handmade quality of their activities. DIY artifacts often retain traces of how they were built. They have rough, unfinished components, with tool marks or the handmade process still visible in the finished object. Because DIYers enjoy hands-on experimentation and the use of available tools, their work often has a low-fi aesthetic that reveals process, unique use of tools, or even failed attempts. For this reason, DIY culture often celebrates nonstandard approaches to craft. For example, folk art often has human figures or scenes drawn with nonphotorealistic features or perspectives. Punk ear piercings or clothing modifications are done with a safety pin, not a professional tool. Similarly, DIY zines are often created at home, using inexpensive tools like a coin-operated photocopier instead of hiring a professional printing house (figure 0.2.2). In this sense, DIY prioritizes material engagement over conversation or conventional institutions. This is why the bricolage of DIY practices skirt across or ignore formal disciplines. John Jordan, in "The Art of Necessity," sees DIY as outsider art. By this, he means that its process works outside of the language system of art and the conventional disciplines of knowledge, practice, and power.[11]

Defining DIY: Contexts of "It"

One appealing aspect of DIY is that the object or activity can be nearly anything: music, haircuts, clothing, videos, automobile repair, or electronics. In other words, "do-it-yourself" uses a generic "it" as a stand-in for any possible thing. Through this, DIY activities can use ordinary objects and materials to sculpt bespoke ones. DIY practitioners also tend to use available materials to shape their own identities or cultures by constructing what they feel is missing from the mainstream.[12] Amy Spencer argues that the "it" of DIY is primarily built as an extension of one's identity or to address something lacking in culture at large.

For the most part, the "it" of DIY can be thought of as a bricolage of whatever is available, and this randomness is part of its creative energy. Claude Lévi-Strauss sees that the rule of bricolage is to always "make do with 'whatever is at hand,' that is to say with a set of tools and materials which is always finite and is also heterogeneous because what it contains bears no relation to the current project, or indeed to any particular project."[13] Reflecting this idea, DIY materials are often everyday items initially designed for another purpose. Lévi-Strauss's bricoleurs, like DIYers, are adept at repurposing odd items into use for an array of diverse projects. While the common definition of "making" has been critiqued for reviving masculine mythologies of technical prowess, I rely on a broader definition of "it" in DIY. This

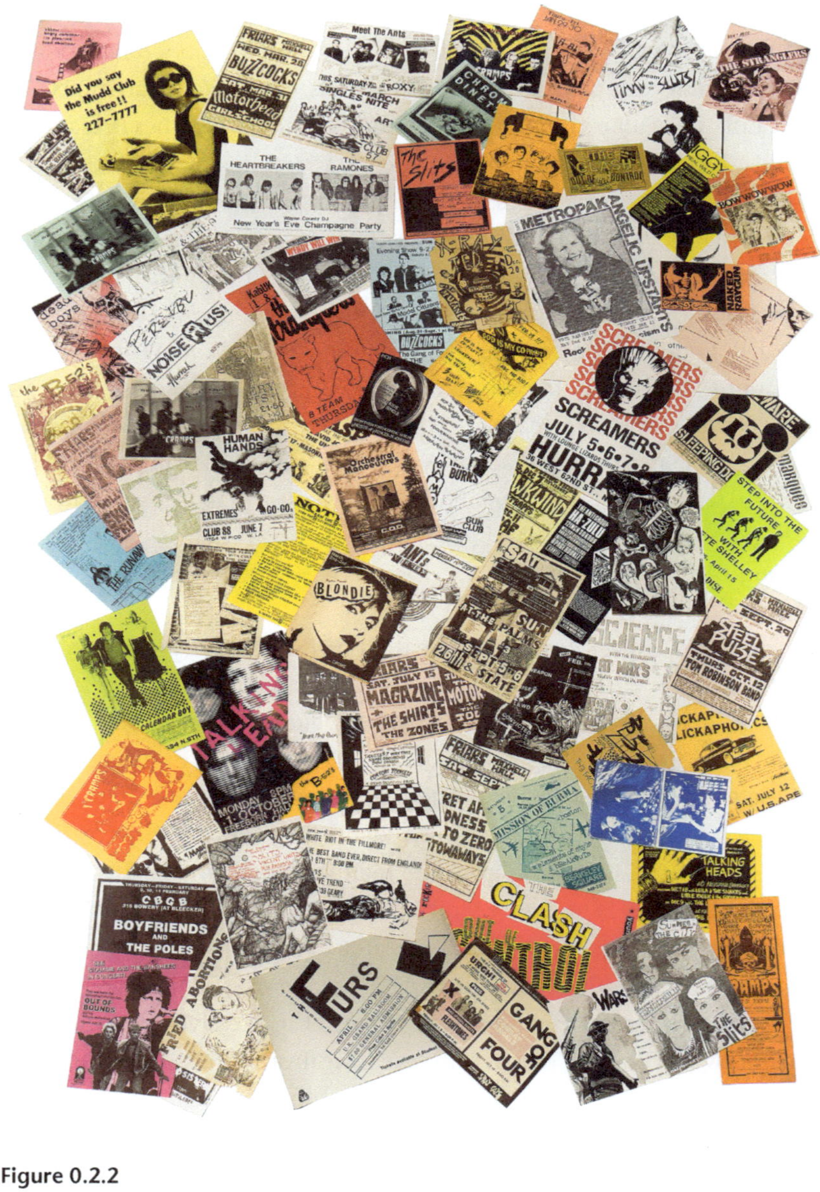

Figure 0.2.2

Punk posters at the Cranbrook Art Museum in the 2018 exhibition "Too Fast to Live, Too Young to Die: Punk Graphics, 1976–1986." Photo PD Rearick, courtesy Andrew Blauvelt, Cranbrook Art Museum.

semiotic ambiguity invites us to explore how people in non-Western contexts approach their craft. The skill of these craftspeople comes forth when they forcefully bend and frugally remix artifacts in unexpected ways.[14]

One example is how everyday materials like plastic drink bottles can be creatively reconfigured into a tool for connecting a water tap to a pipe (figure 0.2.3). Similar "hacks" include modifying and using a plastic drink bottle for dozens of different tasks, including crafting it into a funnel, a soap dish, bowling pins, bird feeders, a broom, a gas mask or a wasp trap.[15] The same approach of frugally reusing objects in novel ways also applies to more complex technologies as well: in northern rural India, "jugaad" vehicles are regularly built using assorted scrap parts.

As a result of using available materials, DIY tends to be inviting and participatory. Spencer cites the skiffle bands of the 1950s as an example of using everyday objects—like washboards and tea chests—to make musical instruments. Everyday materials promote a participatory attitude where anyone can join in and do it themselves. When commonplace objects are matched with a low-fi aesthetic that has traces of the builder's process, it

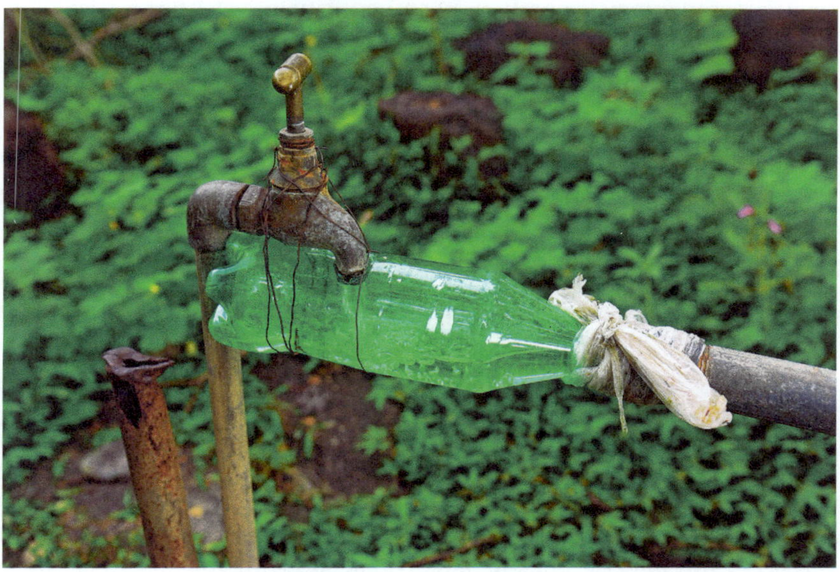

Figure 0.2.3
An example of a jugaad approach to DIY, where a plastic water bottle is used in a frugal and unconventional way to connect a tap to a pipe in Mumbai, India. Photo Dinodia / Alamy.

acts like a visual public tutorial that invites others to do the same. Participation, as a goal, can be idealistic. While I argue that there is a benefit to utopianism, barriers often exist to achieving full participation, particularly when "making" is naively deployed without giving attention to why participants might find its western masculine underpinnings exclusionary. For this reason, in this book I carve out space for international DIY electronic artists who use everyday objects like electronic children's toys as a starting point for their audio or art projects. If you can then become familiar with their work and processes, everyday objects can be an invitation to do the same hack yourself: an old robotic toy dog spotted at a garage sale, for example, can be an invitation to transform the device into a customized pollution-sniffing robot in the style of Natalie Jeremijenko's *Feral Robotic Dogs*.[16] Because DIY relies on everyday materials, its invitation to participate is interwoven through the ordinary technologies all around us.

It: The Authenticity of the Everyday Object

In addition to being an invitation for participation, the use of everyday materials can integrate questions of authenticity into artistic practice. Writing about the emergence of Dadaism in 1934, Walter Benjamin saw Dadaists' use of materials from daily life as having the potential to shock the art world and challenge the larger question of authenticity in that sphere. He wrote that "the revolutionary strength of Dadaism consisted in the testing of art for its authenticity. Still lifes put together from tickets, spools of thread, and cigarette butts were linked with artistic elements. They put the whole piece in a frame. And they show the public: look, your picture frame ruptures the age, the tiniest authentic fragment of daily life says more than paintings."[17] Using everyday and available materials can also question the established practices or knowledge within a discipline. In the framework of this text, electronic projects from the fields of electronic art and experimental design push us to question the role of electronic objects in everyday life. In the case of electronics, Jeremijenko's use of old secondhand toys initially asks us: "Why are we throwing all of this stuff out?" or "Why do I need to learn how to properly design a circuit board when I can get more interesting results by modifying the guts of a $50 toy?" In this book I argue that obsolete technologies are much more than e-waste—they can operate as artistic agents of personal, social, technical, and political change.

It: Materials Pushing Back and Object-Oriented Ontology

Material objects also clearly push back against the wishes of a maker, artist, or tradesperson. Acrylic paint behaves in a certain way on canvas, and metal behaves in a certain way when welded. Amateurs with less training simply experience more of an improvised process because they do not yet understand the "voice" of the materials. In other words, each material has its own affordances, strengths, and weaknesses. Within DIY practice there is a close link between human and technical-material actants, and bricolage disrupts the structures of how objects are conventionally used. DIY practice is a zone where materials and technologies are not just controlled or mastered, but where they exert creativity on the development process. The voice of the materials is amplified by an amateur DIY practitioner's lack of skill when understanding and controlling the tools, methods, or processes of their project. For this reason, a finished DIY project often looks like a battle between the original materials used and the desires of the maker, with no clear winner.

One framework to understand this interplay is Object-Oriented Ontology (OOO), which envisions objects as having their own voices and motivations.[18] Ian Bogost describes OOO as putting things at the center of thinking about existence. "Its proponents contend that nothing has special status, but that everything exists equally—plumbers, cotton, bonobos, DVD players, and sandstone, for example."[19] Similarly, I imagine DIY as an inquiry into the nature of existence, as a territory where the wills of objects often speak loudly enough to be actants in the creative process. The commonplace "it" materials of DIY can have "efficacy, can do things, [have] sufficient coherence to make a difference, produce effects, [and] alter the course of events."[20] In this sense, DIY culture challenges the way we see "power," "efficacy," "coherence," and other attributes as exclusively human traits.[21]

Defining DIY: Contexts of "Yourself"

The third component term of DIY is "yourself." This requires elaboration, since it does not simply mean "alone." "Yourself" in this context is more akin to working independently, outside of the control of a manager. The independence is not isolation from people—it is freedom from managers who influence a project. Accordingly, "yourself" often implies an amateur driven by personal goals rather than a desire for financial gain. DIY inherits both of these connotations, of appreciating amateurism and freeing oneself from managerial constraints.

Yourself: The Hedonistic Amateur

A starting point for thinking about amateurism is thinking about an "amateur" as someone that does a task without having making money as their primary goal. However, the concept of the amateur is significantly more complex than just being "nonprofessional." As the notion of implicit rewards explored earlier suggests, the amateur is an enthusiast driven more by interest and love than by finances, duty, or occupation. This meaning reflects the origin of the term amateur, coming from the Latin *amare*, literally "to love." Historian Rachel Maines describes the inverse of utility-oriented production as "hedonized production." This occurs when the usefulness of a thing is overshadowed by the pleasure in or love of producing it. Maines tracks the transition of utilitarian chores into leisurely tasks.[22] Consider the examples of gardening, hunting, cooking, needlework, home mechanics, and brewing. In these cases, a lack of material hardship assists in do-it-yourself tasks becoming enjoyable hobbies. In times of prosperity and leisure, chores shift into artisanal crafts. This applies to the bulk of DIY production in art: it is a hedonized, pleasure-oriented process. Its pleasure in production is an intrinsic reward.

However, DIY practitioners are not always unified in not being primarily motivated by finances. DIY practices are often done out of necessity and a lack of resources, while some DIY activities can be embarked on as a deliberate choice. Although financial constraint is a significant factor—and a topic I more thoroughly address in its own series of chapters—it is worth emphasizing that DIY practices operate without a commercially motivated endpoint. In other words, finances often *force* a DIY approach, but financial gain is not usually the goal. The mirror is not duct taped to be resold, it is duct taped to be fixed. Similarly, if someone produces a self-published, photocopied zine on their favorite topic, copies are more likely to be given away to friends than turned into a widely distributed commercial publication.

Lisa Gitelman brings clarity to amateurism in her outline of a history of the amateur press. She writes that, "it would be a mistake to define 'amateur' in contrast to 'professional' and leave it at that . . . Again and again, amateurs insist to their readers how hard they work and how much time and effort their papers require, while they also stress that their labor is self improving yet money losing, not profit taking."[23] In other words, DIY production is impacted by the commercial forces of the world, but its practitioners are generally disinterested in exploiting commercialism, independent of whether they are motivated by utilitarian or hedonized goals.

Yourself: Unmanaged, Escaping Taylorism

In relation to not being motivated by financial gain, I argue that the distinction of DIY is in individuals operating outside of the control of an external manager. Florian Cramer states that "DIY thus only exists through its other: an industrial economy."[24] An individual's actions have no oversight by a corporate administrator with an MBA, for example. This applies to both utilitarian and hedonistic DIY practices and is vital to understanding how the "yourself" of DIY is significantly different from just doing something "alone."

The theory of Taylorism is useful here because of its significant impact on contemporary industrial culture. DIY practice vigorously rejects the tenets of Taylorism and its discipline of scientific management. As a result, DIY can be thought of as a design process that can be included in Anthony Dunne's concept of the post-optimal.

Frederick Winslow Taylor, born in 1856, was one of the first promotors of measured and managed efficiency in manufacturing, which was initially called the "task system" and over time became known as "scientific management." This system was based on his extensive observations and workflow tests at the Midvale Steel Company in Philadelphia and the Bethlehem Steel Works, also in Pennsylvania. In the spirit of Lillian and Frank Gilbreth, who were researching efficiency of motion and ergonomics in bricklaying and construction, Taylor analyzed the relationship between time and profitability. He did this by optimizing factory materials, tools, employees, and work arrangements in industrial manufacturing. Taylor's major contribution of relevance to the topic of DIY work was not just the optimization of manufacturing processes. Taylorism was revolutionary in its proposal of a rigid separation between management and manufacturing. Taylor vigorously promoted a split between mental planning and manual production, which sped the transition from craft-oriented apprenticeships into production-line manufacturing in the twentieth century.

Taylor's most noteworthy contributions were published in 1911 as *The Principles of Scientific Management,* which was voted as "the most influential book on management ever published" by the Fellows of the Academy of Management in 2001.[25] Taylorism's antithesis was craft-based production, which he believed placed too much agency and control in the hands of the maker. Taylor's approach was to centralize the scattered and personal understandings of how to craft an object through dedicated managers who scientifically analyzed, optimized, and forcefully standardized production methods. Taylor writes, "it is only through *enforced* standardization of methods, *enforced* adoption of the best implements and working conditions, and *enforced* cooperation that this faster work can be assured. And the duty of enforcing the adoption of standards and of enforcing this cooperation rests with the *management* alone."[26]

This enforced standardization saw significant productivity gains and profitability for manufacturers. Consequently, it laid the path for the fields of business administration and industrial engineering. It also initially resulted in significantly increased wages for workers at companies that had bought into his new, high-productivity mode of scientifically managed manufacturing.[27] Workers saw significant gains in salaries, but this corresponded with a drastic increase in workload. In Taylor's example of Bethlehem Steel Works (figure 0.2.4), workers received a 63 percent pay increase, but the average ore that a worker had to shovel per day changed from 16 tons to 59 tons per person, almost quadrupling individual workload to an output of 369 percent.[28] This staggering increase in output was accomplished by replacing rule-of-thumb methods with scientifically measured and optimized efficiency, by consolidating the training of employees, and by breaking down jobs into discrete tasks.

The cost to society, however, was a growing gap between workers and the items they produced. In essence, personal pride of craft and mental engagement with work in factories were decimated. Staffing was also difficult for companies that embraced the scientific management system with workers performing redundant tasks: "In 1913, for example, Ford was forced to hire more than 52,000 workers to sustain a workforce of about 14,000"—which seems to indicate that most workers did not think that the increase in pay was sufficient for them to endure mind-numbing tasks on an assembly line.[29]

	Old Plan	New Plan Task Work
The number of yard laborers was reduced from between	400 & 600 down to about	140
Average number of tons per man per day	16	59
Average earnings per man per day	$1.15	$1.88
Average cost of handling a ton of 2240 lbs.	$0.072	$0.033

Figure 0.2.4
Taylor outlining the financial operations of Bethlehem Steel Works, comparing the previous staffing, productivity, earnings, and costs to using his task-oriented plan. This saw a drastic reduction in laborers, a strong increase in workload per individual worker, a significant increase in pay, and a significant reduction of company operating costs. *Source*: Taylor 1919.

The previous craft model for work—which was more human-scale and dependent on personal skill and experience—was overtaken by managerial rigor. Taylor's concepts helped form the first Masters of Business Administration (MBA) program in the world at Harvard University, where he lectured.

It is worth emphasizing that the "yourself" component of "do-it-yourself" is predominantly a rejection of being externally managed and optimized. The autonomy of the DIY maker is more a rejection of Taylorism than a social isolation from other individuals. It rejects the "separation between planning and execution which seems to be in our day a common denominator linking all industrial societies together."[30] This autonomy applies to both utilitarian and hedonized DIY practices, and it results in production that is holistic in the spirit of craft-based production and not broken into discrete tasks like an assembly line. As a part of not being managed, the onus is on the individual to figure out on their own what to do. DIY practice is autotelic and self-driven, and depending on the amount of patience and time devoted to the process, the end result can either be a duct-taped rearview mirror or an intricate craft constructed from found materials.

Doing It Yourself: DIY as Post-Optimal Practice

If Taylorism and its efficiency-oriented production lines and management structures are considered a quest for scientifically determining the *optimal* manufacturing process, DIY practice is definitely *nonoptimal*. Taylor abhorred artisanal production. He would have considered DIY production grossly wasteful of time and full of "awkward, inefficient, or ill-directed movements."[31] In other words, DIY is not concerned with scientifically optimizing objects to be manufactured in the most efficient or profitable mode possible. Rather, it puts planning (or lack of planning) back into the hands of the builder. DIY as a nonoptimized mode of production also relates to the concept of the "post-optimal object," a term carved out in 2005 by Anthony Dunne within the context of industrial product design.[32] The post-optimal object, as understood by Dunne, is focused on the user's experience with the finished object over the process by which it is produced. It is worth mentioning that this stands in contrast to DIY work often being a process-oriented hedonistic type of production.

Post-optimal objects start with the assumption that consumer culture has generally reached a level where commercial products are available to most people at low prices, with sufficient features and relative durability. Note that the post-optimal perspective assumes a relatively affluent social and material environment and generally comes from the perspective of product design. In other words, in middle- to upper-class environments many people can fulfill their basic needs for commercial products by going to a Walmart, Target, Canadian Tire, Argos, or other large retailer of commercial goods. Dunne refers to Peter Dormer's description of the problem as follows:

> This is what differentiates the 1980s from 1890, 1909, and even 1949—the ability of industrial design and manufacturers to deliver goods that cannot be bettered, however much money you possess. . . . The rich cannot buy a better camera, home computer, tea kettle, television or video recorder than you or I. What they can do, and what sophisticated retailers do, is add unnecessary "stuff" to the object. You can have your camera gold plated.[33]

In other words, in well-to-do environments Taylor's vision of widespread manufacturing efficiency has already come to pass. Relatively inexpensive and mass-produced items currently represent the most functionally advanced products available. The post-optimal can also be thought of in a nonaffluent sense as an acknowledgment that many aspects of modern consumer technologies do not necessarily bring happiness. In other words, happiness is not directly proportional to how new, fast, or advanced the products you buy are. A more "hipster" perspective might see the rise of artisanal products in the 2010s as an aspect of post-optimalism.

Dunne's argument is that design should move beyond optimizing products toward building objects that improve life experience by being more interesting, poetic, and engaging. I agree. Operating more in the style of a provocative film, Dunne outlines trends in industrial and electronic design where products take the instigative role of asking questions like "What does it mean to be human?" or "What role do electronic objects have in our environment?" These questions transform our perception and consciousness of our surroundings. Examples that Dunne describes include several projects that aim to visualize how radio waves permeate our contemporary environment—like their *Tuneable Cities* project proposal that uses a car and radio scanner to discover how devices like audio baby monitors broadcast private conversations into the public environment.[34]

In other words, Dunne's post-optimal objects are more concerned with asking questions like "How do technologies expose private spaces into the public realm?" than solving functional problems like "How can this product operate more efficiently?" Post-optimal objects leave problem solving to mass-produced products and instead work to uncover a poetic dimension of new experiences for the user. "Driven by poetry, imagination, and intuition rather than reason and logic, they have their own sense, an alternative to our everyday scientific-industrial one."[35]

The post-optimal object can be defined as a designed object that is not reducible to a set of objective metrics like efficiency, speed, resolution, or capacity. Instead, it is culturally or psychologically provocative. As Dunne elaborates, "If user-friendliness characterizes the relationship between the user and the optimal object, user-unfriendliness then, a form of gentle provocation, could characterize the post-optimal object."[36] Dunne's post-optimal object is provocative to the user and observer. DIY goes beyond being post-optimal in its non-Taylorist development process. DIY practice rejects, or does not attain, a mass production mode of efficient productivity. If standard product development is thought of as a quest for optimization, DIY production is both a pre-optimal craft and a post-optimal exploration. *Pre-optimal* refers to a historical mode of craft-based production, and *post-optimal* refers to a response against consumer manufacturing.

DIY production is both pre- and post-optimal in a similar sense to how the Arts and Crafts movement was pre- and post-optimal in mid-nineteenth-century England. John Ruskin and William Morris, central figures to founding the Arts and Crafts movement, saw Victorian-era industrialization as a negative force that degraded workers, social structures, and culture in general.[37] As a reaction against industrial factory manufacturing, the Arts and Crafts movement revitalized preindustrial artisanal production, marrying it with socialist ideals in an attempt to bring labor, products, and capital together for a better quality of life.[38]

Although DIY is a general concept and Arts and Crafts is a canonized "movement" and period in design, both reject mass manufacturing. In cases where DIY is done voluntarily and out of enjoyment for the process, it often takes on a therapeutic role for the maker that echoes the original intentions of the Arts and Crafts movement: it works to reconnect individuals to material craft. But instead of just being disenchanted by industrialization, contemporary individuals also seek refuge from screen-based forms of work and the feeling of being disconnected from how objects are actually made in a globalized consumer culture with "a dark aura of sweatshops, child labor, wage exploitation, pollution, and other dubious specters of globalized production."[39] The globally optimized product is a personally disconnected product, and the tendency toward DIY practice can be seen as a post-optimal attempt to personally reconnect with how physical things are made.

Conclusion: A Definition of DIY

In summary, DIY can be defined as a materially oriented, embodied practice that is individually directed and non-managed. DIY practitioners generally believe there is a value in manual labor, while frequently rejecting the optimized structure of mass manufacturing. This work is often intrinsically rewarding. "Doing" is an embodied activity where hands and mind work together to manipulate physical materials. The built objects often bear the marks of nonstandard, manual, and nonprofessional approaches to building artifacts. In other words, DIY artifacts frequently have a low-fidelity, "folk" look to them. This often is a byproduct of DIY builders making do with whatever is at hand, a bricolage of limited materials and skills. DIY work is typically done by amateurs driven by a lack of resources and the love of making things, and as a result, DIY projects often bear the visual marks of how they were built or cobbled together—and the process itself is important.

DIY projects are generally built using everyday and available materials and are not-for-profit. DIY is driven by an immediate functional need: to fix something or create an object that addresses what is missing from popular culture. Typically, common materials are used in DIY practice. This leads to an aesthetic of openness—inexpensive materials often promote an attitude where others are invited to do it themselves. The challenge of building the object also leaves traces of how it was built and acts as a visual guide for how to build it. "Yourself" implies an amateur that driven by personal goals—part of a search for authenticity—rather than financial gain. As a result, DIY can be thought of as appreciating amateurism and as an effort to break out of managerial constraints. Rejecting standard metrics of efficiency, speed, resolution, or capacity means that these projects can share the attributes of both post-optimal objects and craft production.[40]

In the sections ahead, this book explores and explains specific trends, themes, and ideas that flow through the diverse field of technology-oriented art during the past century, from well-known projects to lesser-known ones. Five core themes emerge from this historical blend of DIY projects in art, design, craft, and technology: work dealing with constraints, "unblackboxed" projects motivated by making technologies more understandable and operationally transparent, work that is concerned with embodied existence, projects done as a form of tactical protest, and projects that are co-opted by industry. The five themes of constraint, unblackboxing, personalization, disobedience and co-option are fleshed out in the chapters ahead with an emphasis on extracting several points of leverage that are particular to the dynamic field of amateur-oriented, do-it-yourself technology production. In the end, we will learn several useful lessons: about how artists can help society; how individuals can powerfully enact political or personal change; and how technological innovation emerges from gaps between traditional disciplines.

Theme 1 Frugal Innovations

Frugality and DIY Electronics: A Thematic Overview

Constraints are ubiquitous in DIY production. Limited resources, money, or skills are commonplace when building anything on your own. This section of chapters focuses on how DIY projects are often driven by constraints, either as an externally imposed financial necessity or as a self-imposed aesthetic choice. Lack of money often forces people to make things on their own or to cobble together makeshift solutions. At other points, individuals can electively choose to embrace constraints—like limiting oneself to a particular material or method—as a part of the creative process. Together, constraints constantly limit how technologies are built, deployed, and used.

Three case studies form the backbone of the chapters ahead: projects include the *Demanufacturing Machine* (1978–1979) by Survival Research Laboratories, *Telephonic Arm-Wrestling* (1986) by Doug Back and Norm White, and the *Toaster Project* (2009) by Thomas Thwaites. The production method, materials, and context of each project are outlined, and the implications of the projects are expanded on to illustrate how constraints more widely shape do-it-yourself practices.

This section starts off with chapter 1.1, "Frugality and the *Demanufacturing Machine*: Zombie Technology, Bricolage, and Hype Cycles." This chapter explores how the group Survival Research Laboratories (SRL) revives dead, discarded, and broken technologies for alternate use in public performance. Expanding on the concept of "zombie media" that Jussi Parikka and I coined in 2012, I argue that these reborn "zombie" technologies help us rethink our relationship to obsolescence, DIY practices, and the way we view art that uses technology.[1] Similar in spirit to Duchamp's "ready-mades," constrained DIY turns the trailing and obsolete edge of technology—visible in end-of-life commercial objects—into a site of creative practice where new artistic life is breathed into dead technologies. This is not simply financial need or aesthetic choice, but a combination of the two that leads to a *Mad Max* style. This chapter tracks the vibrant origins of Survival Research Laboratories by analyzing their first device, the Demanufacturing Machine, which was largely built out of industrial surplus taken from an abandoned factory in San Francisco. The topic of technological obsolescence is examined to highlight how many DIY practitioners use discarded, old, and outmoded materials and technologies in their work.

The next chapter focuses on the most common constraint encountered while building things—a lack of money and resources. This widespread constraint of means is often at the core of why individuals build things on their own: DIY production is often driven by a lack of funding, and "making do" involves cobbling together technologies from affordable tidbits and one's own knowledge. This is a common thread that unites folk artifacts made in Russia,[2] homemade inventions built by Cubans under US economic sanctions,[3] and custom-built farm machinery in rural Saskatchewan.[4] In these cases, DIY is not a stylistic choice but is driven by economic necessity. Building a working thing with few resources demands creative workarounds when finances are lacking, often resulting in what architectural critic Charles Jencks terms "adhocism"—using the resources at hand to solve problems.[5]

Chapter 1.2, titled "Frugality and *Telephonic Arm-Wrestling*: Jugaad, Finances, and Function" explores Doug Back and Norm White's 1986 project titled *Telephonic Arm-Wrestling* as an example of do-it-yourself production that happens as a result of limited finances and resources. The project is an electromechanical system built to enable cross-continental arm wrestling sessions—in other words, people using the device can arm wrestle each other in distant locations. However, the device was built from hardware store torque wrenches, surplus motors from old computer printers, parts that a friend welded, and some custom electronics. I argue that this project represents a type of jugaad, a Hindi term for make-do solutions under severe constraints. Put differently, jugaad is an ad hoc approach to do-it-yourself production where builders use grassroots creativity to cope with limited resources. The concept of jugaad is an important alternative to more institutionalized development processes like design thinking; clever makeshift fixes are, by my estimation, how the bulk of the world's population *does* design. Design scholar Deepa Butoliya defines jugaad as "a frugal, bottom-up, everyday making practice done with an intention of a workaround when faced with a scarcity of resources."[6] Jugaad is not only a method for artists and makers, but also a larger approach and attitude that is important to comprehend if the Global North is to learn sustainability and intelligent frugality from the Global South.

Although the lack of resources is likely the most common type of restriction faced in DIY production, constraints can also be intentionally imposed on the design process as a stylistic or methodological choice by the maker. There are many shades of gray between constraint-by-necessity and constraint-by-choice. Often, artists and designers will use inexpensive materials as part of an aesthetic style or performance; limiting one's work to a narrow band of materials is a common creative practice. For example, Dan Flavin's repeated use of store-bought fluorescent light bulbs shows how constraint-by-choice can emerge into a distinct aesthetic style.[7] Similarly, Jim Campbell's low resolution light works might be seen as another example of technical constraint as an intentional aesthetic style.[8]

One of the more extreme examples of self-imposed restrictions is explored in chapter 1.3, with the *Toaster Project* by Thomas Thwaites. The chapter, titled "Frugality and the *Toaster Project*: Technical Disorientation, Device Paradigms, and Highlowness," examines the British designer's journey in 2008 and 2009 in attempting to manufacture a consumer-grade toaster appliance completely on his own. During the course of the project, Thwaites attempted to dig his own iron ore, make his own plastic, and source his own nickel for the heating element, mica as a thermal insulator, and copper for the electrical wires and plug. In the process, Thwaites's intentionally highlights the disorientation involved when one tries to do something on one's own, and he reminds us that DIY is a relative term—we constantly use the infrastructures and materials around us and never really do anything completely independently. Constraining his project to raw, malleable resources drew attention to the fetishization of artisanal production and the ways the production of raw materials has been outsourced.

Philosopher Albert Borgmann's concepts of technology are useful to bring light to these dynamics, especially his ideas of device paradigms and disorientation that will be highlighted in the pages ahead. Simply put, when we start along a process of trying to make something ourselves, we convert that thing from something we know, like an easy-to-use device, into something we are focusing on and not taking for granted. In Borgmann's terms, DIY often turns devices into "focal things," but this unraveling of the complexities of an object results in disorientation. To put it differently, the process of unpacking and exploring the interior of a device or process can result in overwhelming disorder. Although Thwaites' project is almost a caricature of the disorientation wrapped up in trying to do something yourself, this activity of getting "under the skin of the slick-looking objects that surround us" is vitally important.[9] Technology is powerful in that it can separate the means of producing an object from its functional use, but focusing intently on an object comes as a double-edged sword: it has the power to remind us of the complexity of the infrastructures we depend on, but it also has the power to completely swallow us in confusion.

Through these three examples we can see that DIY approaches to developing technologies can be considered a more pluralist approach that is more inclusive of how everyday design is done by individuals. This nonexpert and folksy design method is interwoven with constraints of various sorts, which can be considered the antithesis of what design theorist Ezio Manzini might term "big ego design," where design experts have god-like control over how devices are made.[10] Instead, understanding DIY practices—at least for professional product and user experience (UX) designers—is an exercise in increasing one's capacity to listen to how design practice happens in the wild. For art historians, it is a reminder of the substantial technical problems wrapped up in the development of a project. For artists, designers, and makers, it is an important reminder that we all struggle with technology and processes, but that these constraints can also be leveraged into useful thematic tools. In the end, constraints remind us that do-it-yourself work is often more about human endeavor than human achievement, and that we are always already impacted by the infrastructures around us.

1.1 Frugality and the *Demanufacturing Machine*: Zombie Technology, Bricolage, and Hype Cycles

A survey of DIY electronic or robotic art would be incomplete without discussing Survival Research Laboratories, who are commonly known as "SRL." The group has staged large format performances across the globe for four decades and has had a profound impact on the international genre of robotic art and Bay Area counterculture. Although SRL has in-gallery exhibition credits to its name, it primarily operates its own performances outside of galleries. The worldwide influence of SRL's "punk robotics" approach, which infuses tinkering with real danger, can be clearly seen in cultural events like Burning Man, engineering contests like sumo robot competitions, and "robot battle" television. Survival Research Laboratories should be understood as the most innovative robotics lab in the twentieth century inside or outside of industry, vastly out-innovating commercial firms like Boston Dynamics in terms of sheer engineering brilliance. This is a story about how SRL began.

The Punk Robotics of Survival Research Laboratories

Survival Research Labs was founded in 1978 by Mark Pauline as an extreme machine and robotic performance group. Existing under the auspices of a pseudo-corporate name, SRL builds large-scale and powerful machines that violently act together in chaotic, stage-like performances. Their devices set fires, destroy objects, smash each other, and terrorize the audience. Shows are a dark ecosystem of mechanized life-forms, like an industrial and otherworldly habitat for lethal devices. Performances often include the "loudest robot in the world"—a 150-decibel hovercraft that is louder than a cruise missile, a "Pitching Machine" gun that shoots two-by-four boards at 200 miles per hour, and a shockwave cannon that assaults the audience with a deafening explosion loud enough to physically blur one's vision. Added to this is an additional cast of bizarre, large-scale machines welded together from rusted military and industrial surplus, large photographic panels that feature naked bodies and religious symbolism, and actual animal carcasses and parts. You walk away from their legendary shows smelling like diesel, rotting trash, jet fuel, and burning flesh—and outdoor performances of the military-grade equipment typically end by being shut down by local police or fire departments.

The best way to make deeper sense of Survival Research Labs is through the group's founding in 1978. The initial device Mark Pauline built—the *Demanufacturing Machine*—documents the origins of a historically significant cultural organization and provides insight into how constraints can be part of artistic practice (figure 1.1.1). The *Demanufacturing Machine,* although functional as an electromechanical device, operates as postindustrial punk. That is, it shows how constraints by necessity and choice can push artists

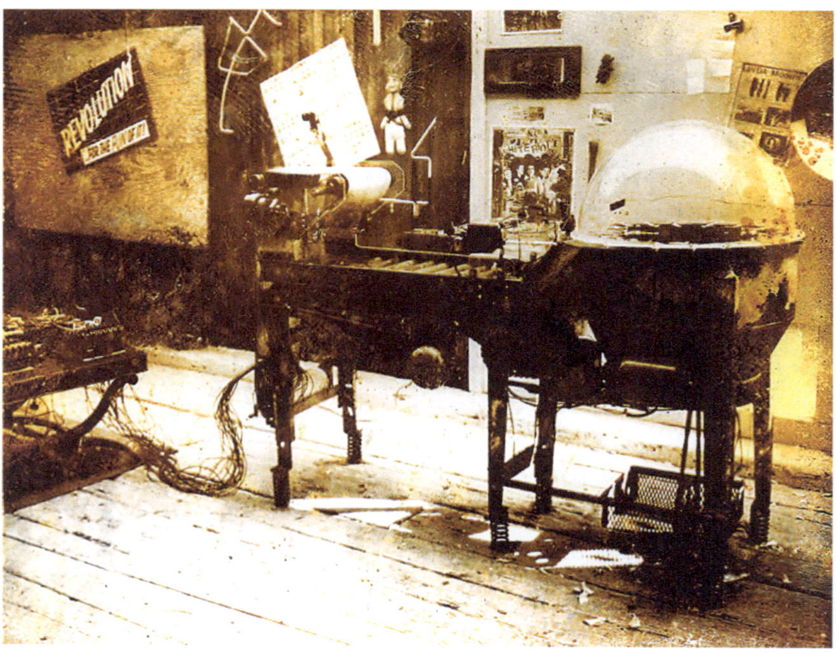

Figure 1.1.1
Archival photo of the *Demanufacturing Machine* (1979) by Survival Research Laboratories. Courtesy Mark Pauline.

into building theatrical forms of technological bricolage by misusing obsolete technologies. If jugaad follows a motto of "make it work," SRL's is "make it spectacular." In particular, SRL revives dead, discarded, or broken technologies into a zombie-like afterlife that in some ways resembles (and predates) the postapocalyptic dystopia of *Mad Max*. I argue these "zombie technologies" help us rethink our relationship to technology, and the way we view its relationship with electronic art.[1]

Starting SRL: Mark Pauline and the Bricolage of the *Demanufacturing Machine*

The idea for building the *Demanufacturing Machine* originated in late 1978 with Mark Pauline—who had graduated the year before from Eckerd College in St. Petersburg, Florida, where he studied with the antiauthoritarian and pacifist "engaged theater" professor James R. Carlson.[2] Around the same time, Pauline had been doing guerrilla billboard modifications in San Francisco, pasting words like "KILL" and "PAIN" to invert the meaning of commercial advertisements for products like cigarettes and liquor. Over the course of a week, Pauline drew up a plan for what he could do for a living, and came up with the concept that would guide the remainder of his life: form an organization to build audacious machines and creative DIY-style robots and do shows with them across the globe. Pauline also had a pastime of breaking into and exploring abandoned buildings. Accordingly, his initial concept was to build things using equipment found or stolen while spelunking through decommissioned warehouses and factories. Pauline settled on the name of Survival Research Laboratories and made the organization "official" by drawing up a display advertisement and placing it in the November 1978 issue of *Boulevards*, a free counterculture-oriented punk tabloid magazine in San Francisco.

Pauline's first effort as SRL was the *Demanufacturing Machine*, built from equipment taken out of the recently decommissioned Falstaff Brewery.[3] The inspiration for the machine, and SRL in general, was to repurpose and invert industrial and military tools. Over the course of a couple of months during the winter of 1978–1979, Pauline bypassed Falstaff guard dogs and security guards to collect pieces of brewery equipment—a conveyor belt, a mixing/chopping mechanism, motors, and a fire alarm bell. He used only equipment rescued from the assembly line as a form of constraint-by-choice. This tendency to use found or stolen materials carries through punk and DIY subcultures: for example, in the 1985-1987 film *Nadie es inocente* ("No one is innocent") by Sarah Minter, young punks featured in the film dig through a municipal landfill outside of Mexico City for discarded clothes that they try on, cut up, tear, and reuse on the spot.[4] In a similar spirit, SRL digs through industrial trash to repurpose it.

The core of the *Demanufacturing Machine* resembles a hip-height, rusted, long metal table with six legs. A drum-like wall against one end is crowned with a large, clear plexiglass, bubble-like dome that rises up to shoulder height. The top of the table surface features a conveyor belt that feeds objects through a plexiglass door and into the dome, which houses a lawnmower-type spinning blade. Directly underneath the blade is a metal mesh basket that catches shredded bits of objects. The basket, connected by a belt to an electrical motor, periodically flings its contents onto the floor to the side of the machine. At the opposite end of the machine is an additional conveyor belt that resembles a vertical scroll that can be used to load up objects to be demanufactured. At the center of the device, mounted underneath the main conveyor belt, is a mechanical fire alarm bell with a flashing red light. It features no glossy casing; its high voltage power supplies, gears, chains, pulleys, and belts are completely exposed. Its rusted frame and violently moving parts threaten injury or tetanus. The machine visually resembles a cross between a mad inventor's hack job and a postapocalyptic movie prop.

When in operation, the mechanical fire alarm bell rings loudly and the light flashes. A few seconds later, the main conveyor belt starts and the blade spins. It takes about four seconds for an object to travel the length of the conveyor belt toward the blades in the drum, where it first passes through a plexiglass door that shuts behind it before the object is shredded. The object explodes as it hits the blades, similar to running over an object with a lawnmower, but the shrapnel blasts up and swirls around the transparent dome before gradually falling into the basket below (figure 1.1.2). Ten seconds later, the metal basket pitches the chewed up bits with a few heaves. After this process has been completed, the alarm and light deactivate and the device goes dormant until it is activated again.

Pauline found military and scientific uses for technologies to be incredibly drab.[5] In response, he built the *Demanufacturing Machine* to

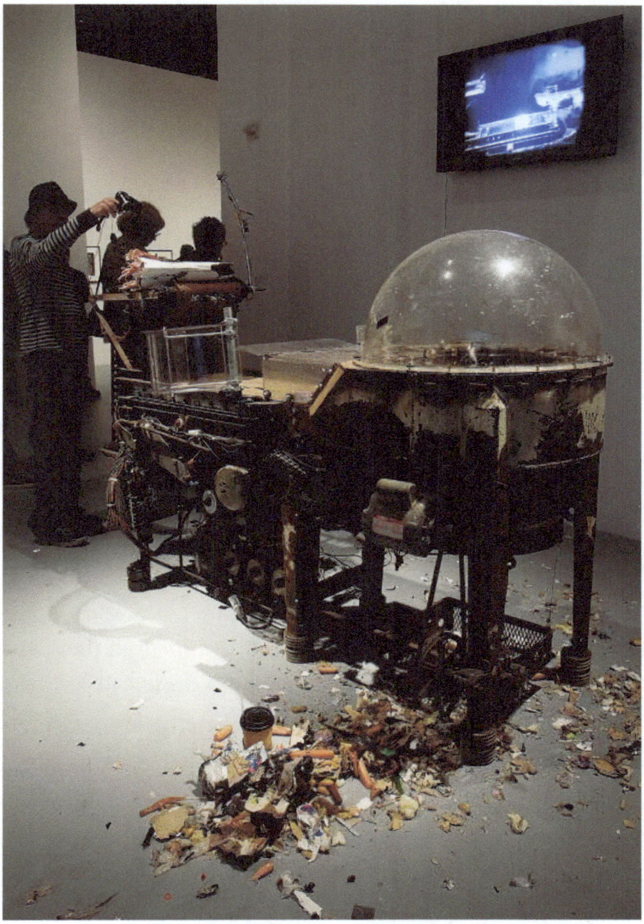

Figure 1.1.2

The *Demanufacturing Machine* at the Geffen Contemporary at the Museum of Contemporary Art, Los Angeles, in 2011. Courtesy Karen Marcelo.

show how these technologies could be poetic, spectacular, and engaging by hacking them into street-level theatrical spectacles. In the process, Pauline drew from the avant-garde "propaganda" skills he had learned at Eckerd College while studying theater as an undergraduate student.[6] He was significantly affected by James R. Carlson, an anti-authoritarian and pacifist professor of "engaged theatre." Pauline's unique twist was to fabricate avant-garde, punk mechanical theater out of military-industrial surplus.

Machine Sex

When selecting a site to unveil the *Demanufacturing Machine*, Pauline gravitated to a highly visible Chevron gas station a block south of Washington Square Park in San Francisco. Adding to its appeal, it was closed on Sundays and had an available electrical plug. The performance was billed under the name "Machine Sex" and occurred on February 25, 1979 (figure 1.1.3). As the machine was being set up, the owner of the gas station appeared and demanded to know what was going on. Pauline appeased the owner by telling him, "Here's twenty bucks . . . just enjoy the show." By this time, artists in the know and passersby had gathered to watch the machine in operation (figure 1.1.4). A loud, cut up soundtrack by The Cure played as the *Demanufacturing Machine* processed dead pigeons wrapped in robes. During the eight-to-ten-minute performance, the fire bell continually rang over the top of the soundtrack as the motorized basket ejected the raw entrails, muscle, and feathers of the birds into the audience. To mark that the show had concluded, 5,000 "Pharaoh's snake" fireworks glued to an eight-by-eight-inch cylinder were set off. This type of firework does not blast like a firecracker, but triggers a chemical reaction that has the appearance of a growing snake. This assembly created thousands of golden-hued snakes growing out of the cylinder mass, like an octopus spontaneously growing a thousand writhing molten arms.

The "machine sex" that was going on was not some form of consensual robotic love: this was more like mechanical disembowelment at a meatpacking plant. This theme of the dominance of technology has stuck with SRL for forty years, reminding observers of the power of technology.[7] SRL uses the shock of mechanical violence to distress viewers out of

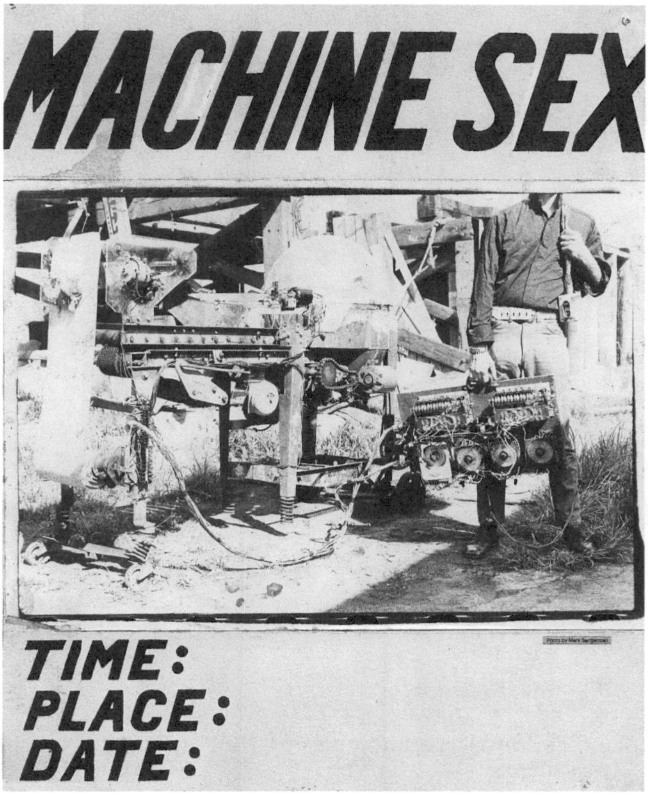

Figure 1.1.3
Poster for the "Machine Sex" event featuring the *Demanufacturing Machine,* which occurred in San Francisco on February 25th, 1979. This was the first machine-based event by Mark Pauline operating under the auspices of Survival Research Laboratories. Courtesy Mark Pauline.

their everyday frame of mind and into an alternate world of machines. Instead of a standardized "extreme technology" exhibit of technological prowess, like an air show, these theatrical happenings, where machines perform according to their own rules, were unnarrated. As a result, they intentionally leave viewers bewildered. The entire performance resembles a nihilistic hedonism that promotes embodied, visceral experience—one that is genuinely dangerous and scary. SRL is not traditionally theoretical or argumentative. Rather, its force comes from using the punk theater of old-fashioned visceral danger to shock us out of existential drabness.[8]

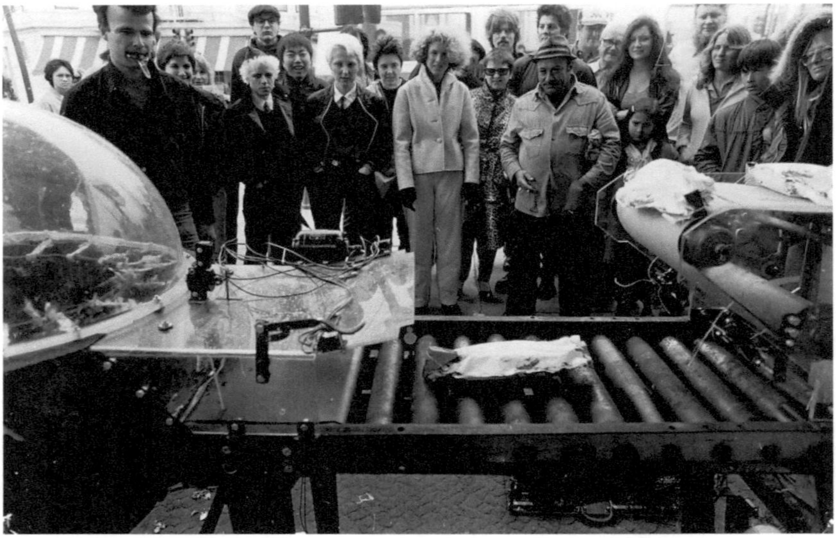

Figure 1.1.4
The *Demanufacturing Machine* processing pigeons during its initial performance at the "Machine Sex" event on February 25, 1979, in San Francisco. Courtesy Mark Pauline.

The Bricolage of Zombie Technologies in DIY Practice

Survival Research Labs highlights the creativity of constraint: obsolete mechanisms sculpted together into provocative public performances. In this sense, it is part of the San Francisco counterculture that was left behind in the city's transformation into a corporate-friendly extension of Silicon Valley. This can be thought of in contrast to the neoliberalist Californian Ideology that portrays high tech as an emancipation from political and geographical limitations.[9] SRL's repurposing and remixing of obsolete technologies—plus the large mechanical scale and overt use of raw violence—is missing in most technology-oriented discussions of new media art and mainstream representations of the maker movement. The zombie-like revival of old mechanisms is reminiscent of the Swiss artist Jean Tinguely, who began experimenting with mechanical movement as early as 1948, using motors to activate sculptures built out of a bricolage of broken apart machines.[10] Tinguely's experimentations similarly reflect an embrace of Dadaist anarchism in an electromechanical format. Acknowledging and embracing the fragility of mechanical devices disrupts the distinction between an admiration of the graceful power of a functioning machine and the cacophony of malfunction. Mark Pauline's group sees constraints as an opportunity for theater, creating jarring mental experiences by breaking down the capitalist relationships that people form with technology. SRL takes clichés, picks them apart, subverts them, and recombines them through performance. In this way, they subversively draw people to question their everyday relationships with technology.

Science fiction author Bruce Sterling described Survival Research Labs as "everything that is repressed in the sterile prison of so-called rational engineering return[ing] in a hideous and terribly authentic guise of claws and spikes and fangs."[11] Their interpretation of post-optimal technology rails against efficient Taylorist factories and the standard-issue "white picket fence" American Dream. Instead, in taking direct material inspiration from machines, they may be thought of as a more pessimistic relative to Japanese Device Art artists like Nobumichi Tosa, Kozuhiko Hachiya, Toshio Iwai, or Ryota Kuwakubo. However, SRL has no interest in commercial product design.[12] From a material and methodological standpoint, they are similar to Reed Ghazala and other circuit benders that take obsolete devices and creatively rewire them into custom-built audio "zombie media" devices by reanimating vintage electronics.[13]

Hype Cycles and the Zombie Afterlife of Industrial Surplus

Jussi Parikka and I refer to zombie media as "concerned with media that is not only out of use, but resurrected to new uses, contexts, and adaptations."[14] This concept is useful in describing how SRL feeds from the heavy industrial-mechanical terrain of San Francisco that was gradually displaced by the information economy.

One visualization to describe the general dynamics of how technologies evolve is Gartner Group's Hype Cycle diagram. This diagram graphs the excitement and sales of emerging technologies for corporate planning of when new products will reach mainstream adoption[15] (figure 1.1.5). In this model—where time runs from left to right and excitement and sales adoption is represented by height—new technologies and products go through a "hype cycle phase" (solid line, left of diagram) that is typified by the excitement over a new technology outstripping its ability to

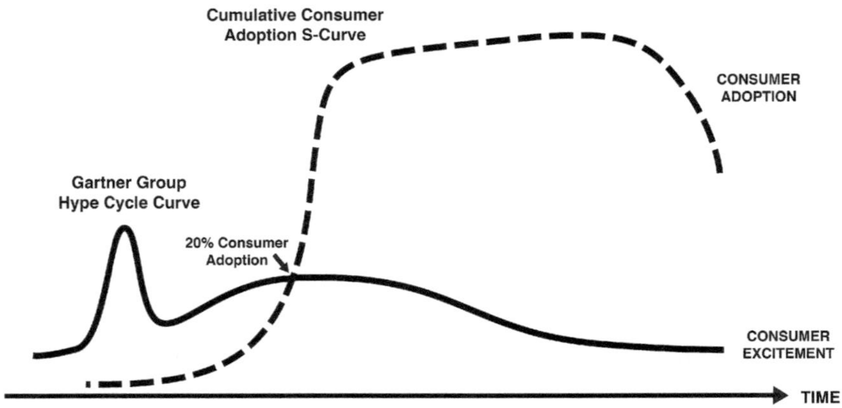

Figure 1.1.5
Gartner Group's Hype Cycle and Adoption Curve diagrams, graphic representations of the economic maturity, adoption, and business application of specific technologies. Image Garnet Hertz.

deliver actual benefit to people. This results in an initial rise and crash of excitement as engineering maturity lags consumer expectations. As engineering and product development improves into usable products, individuals begin purchasing the technology as a commonplace consumer commodity (dashed line). As consumers adopt the technology, sales continue until the product becomes obsolete, when adoption drops off (right of diagram).

This model is typically applied in business contexts to temper the liberation and supersession promised by new technologies, but it neglects several significant factors in material culture. Gartner's hype cycle does not have any capacity to look at how factors like reuse or repurposing play out, for example. Similarly, most artists discussed in this book do not generally follow this curve. Artists working with technology have significantly more interest in either cutting-edge technologies that have not yet reached the mainstream (new media) or trailing edge technologies that are obsolete and free (zombie media)—the excitement curve of interest in technologies is generally not the same with artists as it is with average consumers. Instead of looking at sectors of products from the perspective of consumer functionality, many users are driven by experimental potential and a lack of finances, which results in a graph that is almost the inverse of consumer behavior (figure 1.1.6).

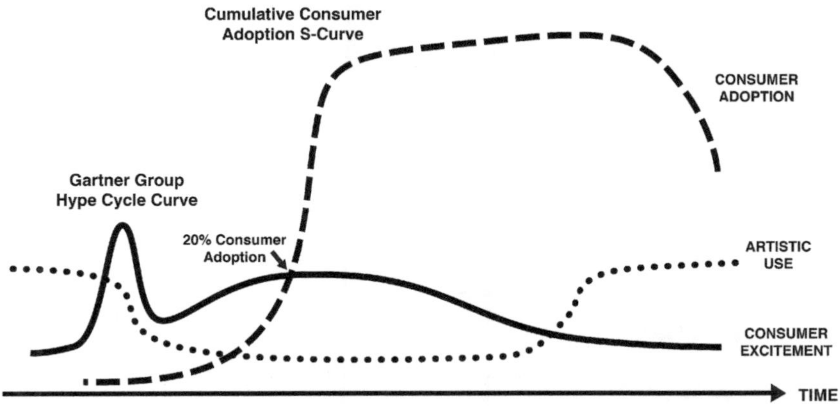

Figure 1.1.6
"Artistic Use" dotted line curve corresponding to bricolage added to the Gartner diagram, showing that artists often are interested in specific technologies before or after they reach mainstream success. Image Garnet Hertz.

Splitting the Gartner Group Hype Cycle diagram into three distinct phases helps in rethinking these dynamics, with the "new media" phase on the left, the "consumer commodity" phase in the center, and the "zombie media" phase on the right (figure 1.1.7). The new media phase involves exploring the potential creative and business uses of the technology, with artists, inventors, and other researchers continually experimenting with an array of new techniques and devices (figure 1.1.7, left). As engineering maturity and consumer adoption (represented together as a dashed line) grows, interest from artists usually wanes, perhaps as a sign that artists typically step aside from working on developing technology objects when they become formal engineering endeavors (figure 1.1.7, center). Put differently, experimental hardware bricolage is followed by hardware engineering, which then reverts to hardware bricolage after the product is obsolete, cheap, or free. Although artists will use all materials and technologies while building work, they and other DIY practitioners have more advantages to the far left or far right of this diagram, primarily because of the lack of money and organized competition. The far right of the graph is particularly interesting, and it is where the *Demanufacturing Machine* might sit if plotted: it used a bricolage of mature technologies that were available (or stolen) at no cost.

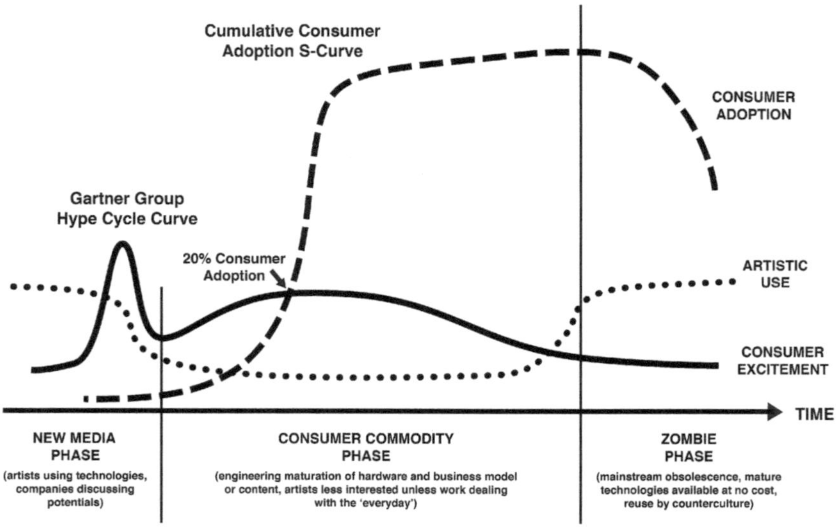

Figure 1.1.7
New Media, Consumer Commodity, and Zombie Phases overlaid on top of the Gartner Hype Cycle and Cumulative Consumer Adoption S-Curve diagrams. Image Garnet Hertz.

The limitations of Gartner's visualization are more apparent when looking at how a "bricoleur" builds things or how a "hipster" repurposes things as nonindustrial forms of making things. A bricoleur, who uses a DIY-style approach to building things, is defined by Claude Lévi-Strauss as the following: "The 'bricoleur' is adept at performing a large number of diverse tasks; but, unlike the engineer, he [*sic*] does not subordinate each of them to the availability of raw materials and tools conceived and procured for the purpose of the project. His [*sic*] universe of instruments is closed, and the rules of his game are always to make do with 'whatever is at hand'. . . . The elements are collected and retained on the principle that 'they may always come in handy.'"[16]

Under these terms, Survival Research Laboratories is more a bricoleur of devices than an engineer of devices: the raw mechanisms that are used are a core part of the inspiration for the final device. In the case of the *Demanufacturing Machine*, the form and function of the Falstaff Brewery equipment still inhabits the final device. It is significantly reconfigured into a new kind of creature, but its industrial origins are still visible and add to its patina of believability. An engineered approach on the other hand, according to

Lévi-Strauss, places the project concept at the center and configures materials and tools around it. This listening to the unique voices of found materials and objects is worth flagging as a unique attribute of DIY and craft practices.

Also, other forms of reuse are not covered in Gartner's visualization, including objects repaired and upgraded, resold, or repurposed. Although the fashionable reuse of objects is an additional topic worthy of its own analysis, the early twenty-first-century hipster trend of fetishizing "genuine" things is also not accounted for in the hype cycle.[17] Neither is the rise of nostalgia for objects approximately twenty to thirty years after they drop out of the mainstream, perhaps like the uptick in interest in vinyl or cassette tapes in the early twenty-first century. If the hype cycle diagram is extended to the right, several technologies would feature a second hype pattern fueled by the nostalgia of middle-aged people recalling their youth.

Hacking the Military-Industrial Complex: Beyond "Toys for Boys"

An ongoing theme in SRL's work has been the misuse and reappropriation of industrial and military technologies into a public performance. SRL works with several of the stereotypes of white masculinity—or at least it clearly plays with a combination of powerful machines, rugged individualism, violence, noise, and aggression. SRL specializes in mechanical spectacle, similar in some senses to a monster truck show, a tractor pull competition, or a demolition derby. Although there is an aggressive and punk aspect to the work, it would be a shortcoming to view it as only a display of masculine machismo. SRL's mission is to redirect the techniques, tools, and tenets of industry, science, and the military away from their typical manifestations—and in the process it has found both innovative technical applications and a diverse array of participants in its work.

SRL was born out of the San Francisco punk scene in the late 1970s, and although the organization and the *Demanufacturing Machine* was Pauline's construction, SRL soon started functioning as a team effort, which included over thirty female collaborators in the mix.[18] The women of SRL are responsible for much of its work, including machining, welding, Internet telerobotics coding, security, mechanics, electrical, carpentry, and video production.

Yes—SRL worked with surplus from the military-industrial complex and many used clichés of straight-white-cis masculinity, but its punk style was primarily concerned with democratizing the spectacle of technology, both with its audiences and through its collaborators. SRL recruited crew of all sorts—a sample newsletter from 1994 reads, "We are accepting volunteer assistance of skilled and nonskilled types. Almost any kind or level of skill is applicable."[19] In this way, SRL echoed the core theme of DIY in punk music: "It was characterized primarily by the fact that anyone could just pick up an instrument and play. . . . Musicians, reacting against the hyper-professionalism of the late sixties and early seventies, distinguished themselves by taking on amateurism as a badge of honor."[20] Instead of using musical instruments, SRL takes a DIY approach through hacking industrial control equipment, high-energy physics devices, military weapon systems, heavy industrial mechanisms, and industrial robots.[21]

And although Survival Research Labs extensively reuses surplus mechanisms and other forms of zombie technologies, it is important to remember that it also uses and invents a plethora of advanced technologies. Old, repurposed scrap is generally mashed up with ultra-high tech, and making both of them work together takes raw engineering talent. Survival Research Laboratories holds a number of engineering firsts to their name, including being the very first group to stream live, wireless video over the web in 1997 in Austin.[22] They were also the first civilians to code and use telerobotics for lethal devices over the web in 1997 (in case that is your kind of thing), they performed the first mechanical test of a robotic swarming/flocking algorithm in 1992, and they were the first to use an animal-controlled robot in a performance.[23]

SRL's work is clearly a violent spectacle that works with industrial and military surplus, but it works to expand these materials into creative applications and to wider audiences. As a part of this, Survival Research Labs reinvents and innovates new technologies, plus it does this with formidable feminine energy. Better and more recent histories of punk subcultures portray a diverse mix of creators and participants adding to the intersectional components of the community—and this is similar in the history of Survival Research Laboratories.[24]

Conclusion: Lessons from Survival Research Labs

The *Demanufacturing Machine* reminds us that artists—as DIY practitioners—build technical objects far outside of the standard model of product development. This type of bricolage—hardware-based remixing that builds off of the past lives of technologies—is constantly being done alongside the standard mode of product engineering. Although new technologies are constantly being developed, on the other side of their life cycle they are reconfigured, rewired, and blended through artistic practice. In de Certeau's terms, "These 'ways of operating' constitute the innumerable practices by means of which users reappropriate the space organized by techniques of sociocultural production."[25] Survival Research Labs is a historically significant "punk robotics" group, and it highlights a vital example of how pieces of obsolete industrial equipment can take on a new technical and artistic afterlife. In the process, we are reminded that there is a significant gap between how sales-oriented visualizations like Gartner's hype cycle only partially represent how material culture works. Old objects never disappear. Rather, they bleed into the present like a palimpsest or find themselves put together in a Frankenstein patchwork like the *Demanufacturing Machine*.

1.2 Frugality and *Telephonic Arm-Wrestling*: Jugaad, Finances, and Function

The fields of electronic art and experimental design and even "proper" disciplines like mechanical engineering have abundant makeshift technical solutions. In other words, people that make things typically all make do with limited resources in some regard. Artists may want to build something complex, but they might only know how to use simple things. Similarly, individuals that are attempting to build or fix something often just hack away at a makeshift solution with what is at hand. Although the process of engineering often approaches design from the mindset of physics, thermodynamics, or mechanics, the practical aspect of innovating often involves grabbing something nearby and hacking away at a solution.

Adhocism and the Frugal Innovation of Jugaad

To use a term coined by the architects Charles Jencks and Nathan Silver, the world of innovation is filled with "adhocism," or an improvised collage of available materials. This improvisational approach incorporates both old and new in unexpected ways, and is a common practice in everyday life and professional practice: adhocism "involves using an available system in a new way to solve a problem quickly and efficiently."[1] It builds off of the Latin term *ad hoc,* which describes when something is fashioned for a specific purpose from whatever is immediately available.

"Jugaad" is a similar concept from India that is an incredibly useful term that helps enrich our concept of ad hoc technical solutions and DIY culture in general. Jugaad—originally from the Hindi word जुगाड़ (jugāṛ) and pronounced "joo-*gaad*"—is a term that generally refers to a lower-class social practice of functional fixes and workarounds using found materials.

Jugaad's closest English translation is similar to the terms "kludge," "MacGyvering," or the British "bodge."[2] The Japanese concept of "urawaza" or the German concept of "Trick 17" is also similar. Some scholars writing on technology describe jugaad as an indigenous form of innovation to India, as a frugal-oriented approach to Design Thinking that has emerged as a mindset in India where scarcity and limited product choices are a part of daily life.[3] Along this line, the term is often portrayed in a positive light as creative improvisation—as a kind of local ingenuity that depends on persistence and spontaneous skill.[4] It can also refer to working around

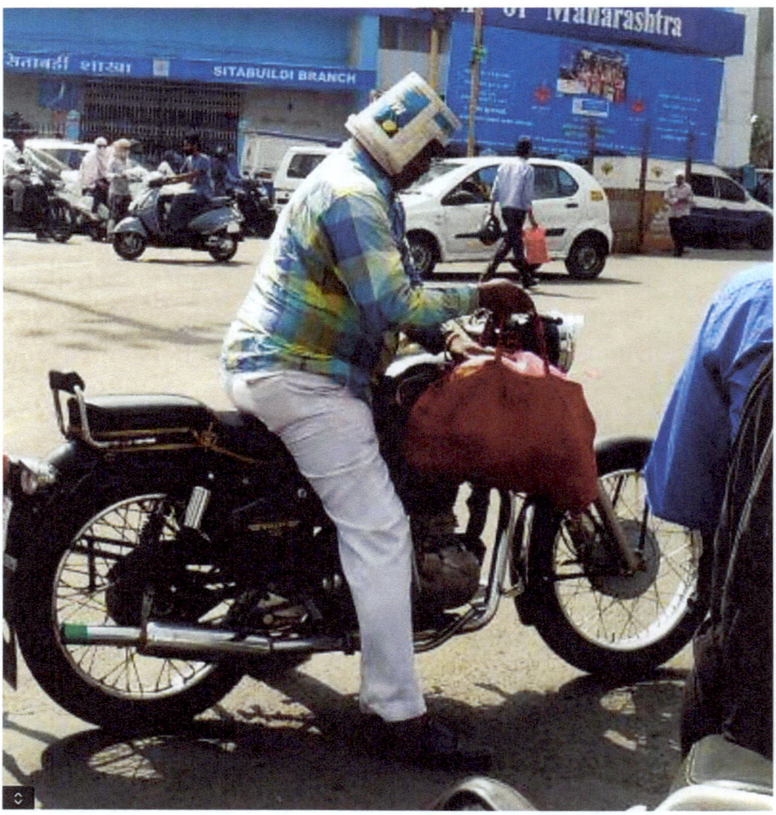

Figure 1.2.1
Homemade jugaad motorcycle helmet made from a modified paint bucket worn on the head in Nagpur, India. This was likely built as a response to a mandatory helmet policy enforced by local police. Courtesy Deepa Butoliya.

formal bureaucracy through illegal activities or corruption, like the informal workaround of paying off officials.[5,6] Criminologist and anthropologist Beatrice Jauregui tracks the core of jugaad to the Indo-European root word *yug*—a union or bringing together of things—which forms the core of modern English terms like yoke, yoga, juxtapose, junction, and junta.[7] In other words, jugaad brings together disparate things in an unexpected and ad hoc way. Jugaad permeates the DIY mindset.

Perhaps the clearest and most succinct definition of the term for our purposes comes from technology scholar Deepa Butoliya, who defines jugaad as "a frugal, bottom-up, everyday making practice done with an intention of a workaround when faced with a scarcity of resources."[8] Butoliya provides examples of a light bulb surrounded by a plastic bag coated in oil to catch insects, plastic spatulas skewered through cabinet handles to keep them closed, or a paint bucket cut into a low cost motorcycle helmet (figure 1.2.1).[9]

Ad Hoc and Jugaad Approaches: Yashas Shetty

The technique of using found materials to build ad hoc solutions can be significantly more elegant than a bucket on someone's head to avoid a ticket, however. Technical improvisation with limited resources is arguably just part of being a contemporary artist. One artist from India that highlights the gracefulness and refinement of DIY-style approaches is Yashas Shetty, founding member of the Center for Experimental Media Arts at the Srishti Institute of Art, Design and Technology in Bangalore (figure 1.2.2). His creative work involves experimental music and sound, and his performances incorporate DIY and homemade instruments made with discarded electronics and other found objects.

Like other contemporary artists working with technology, Shetty's work is a bringing together of disparate academic disciplines, feeding from the sciences, the visual arts, contemporary music, and cultural studies. In his work, he creates assemblages and mashups of various life forms using techniques borrowed from genetic engineering and synthetic biology—and these are presented with the intention of creating dialogue between artists, scientists, and the larger community. Shetty's disciplinary and methodological roots are diverse, and the quality of the work he produces reflects this richness. For example, in 2012 at Gasworks in London he used hacked

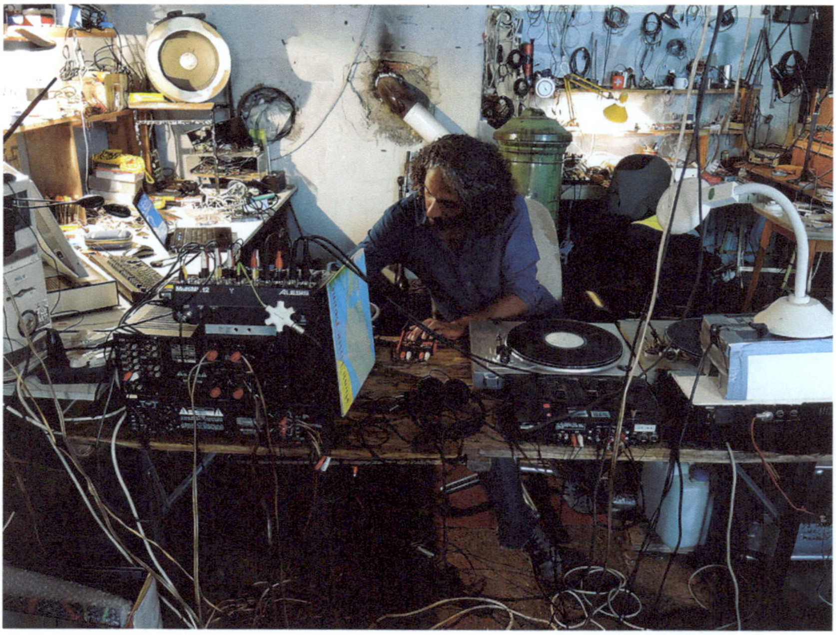

Figure 1.2.2
Yashas Shetty, founding member of the Center for Experimental Media Arts at the Srishti Institute of Art, Design and Technology in Bangalore. Photo Bostjan Leskovsek. Courtesy Yashas Shetty.

consumer electronics found locally to build a self-contained laboratory for hosting public workshops in genetic engineering.[10] Shetty's jugaad-inspired work is innovative in that it intersects art, science, and pedagogy, and creates useful interdisciplinary bridges in the real world. In this case, a jugaad approach of pulling things together applies not only to materials and technology, but also to disciplinary reach.

There are also a number of terms similar to jugaad and adhocism that exist in different cultures. Brazilians refer to this approach as *jeitinho* or *gambiarra,* while in China it is often called *zizhu chuangxin.*[11] Kenyans refer to it as *jua kali* and the French refer to it as *Système D.*[12] Ernesto Oraza, in documenting cobbled together inventions in Cuba, refers to it as "technological disobedience."[13] And no matter what you call it, the core principle is the same: this cluster of approaches includes quick and dirty workarounds that are effective in solving mechanical or technical problems by cleverly using cobbled together parts.

Telephonic Arm-Wrestling and Jugaad: How Financial Constraint Impacts DIY Production

This dynamic of jugaad design in the field of electronic art is also clearly illustrated in the project *Telephonic Arm-Wrestling* (figure 1.2.3) built in 1986 by Canadian artist Doug Back and Canadian American artist Norman White. It is an interactive robotic project that enables two individuals in remote locations to physically arm wrestle each other. The development details of Back and White's project paint a familiar picture of how DIY projects are often scraped together with more creative ingenuity than financial capital.

Telephonic Arm-Wrestling began as an evening conversation between Doug Back, who taught at the Ontario College of Art in Toronto (OCA, now named the Ontario College of Art and Design University), and Carl Hamfelt, a student at the same institution. The concept was for participants in two different physical locations to arm wrestle by transmitting physical movement over telephone lines. The concept started out as a simple contest of physical strength mediated by technology. Back's OCA colleague Norm White suggested that the device could be built as a motorized, force-transmitting system that sent "kinaesthetic information via telephone modem signals . . . to send a tactile sensation over the phone."[14] Initially, Back had conceived of the project as a Cold War peacemaking technology: one could be installed

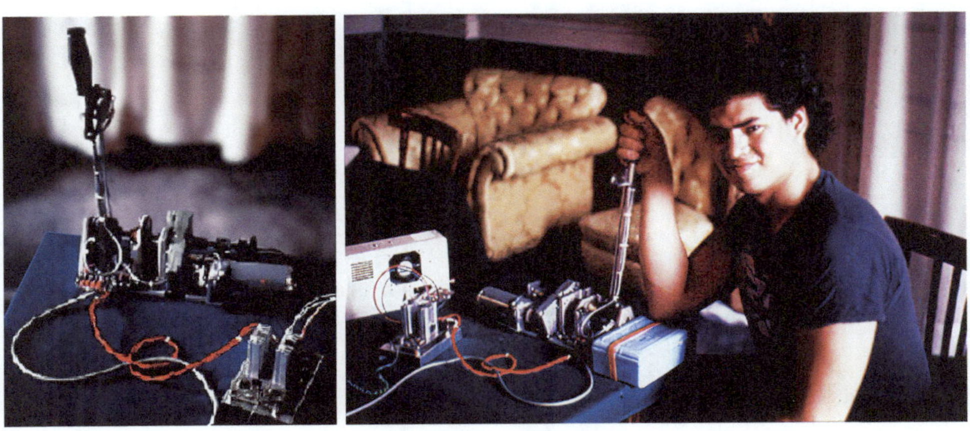

Figure 1.2.3
Technician Ian McGuigan demonstrates the *Telephonic Arm-Wrestling* system by Doug Back and Norm White. Courtesy Norm White.

on President Ronald Reagan's Oval Office desk, while another could be put on Mikhail Gorbachev's desk in the Kremlin—with the two of them using the device to settle any potential disagreements.[15] Remote-controlled arm wrestling was pitched as a kind of humorous symbol of Cold War tensions, rendering it as a clichéd contest of masculinity or physical strength.

The project was then proposed to Derrick de Kerckhove, Director of the McLuhan Program at the University of Toronto, who quickly fell in love with the idea and wanted to exhibit it. De Kerckhove met with the head of engineering at the University of Toronto, who estimated that a professional version built in the Engineering Department would cost $75,000 CAD—significantly too expensive for the artists or de Kerckhove. After sketching out some ideas, Back and White estimated that the system could be built with DIY methods for around $200 by using surplus parts and tools from a local hardware store (figure 1.2.4). A forearm-length torque wrench was

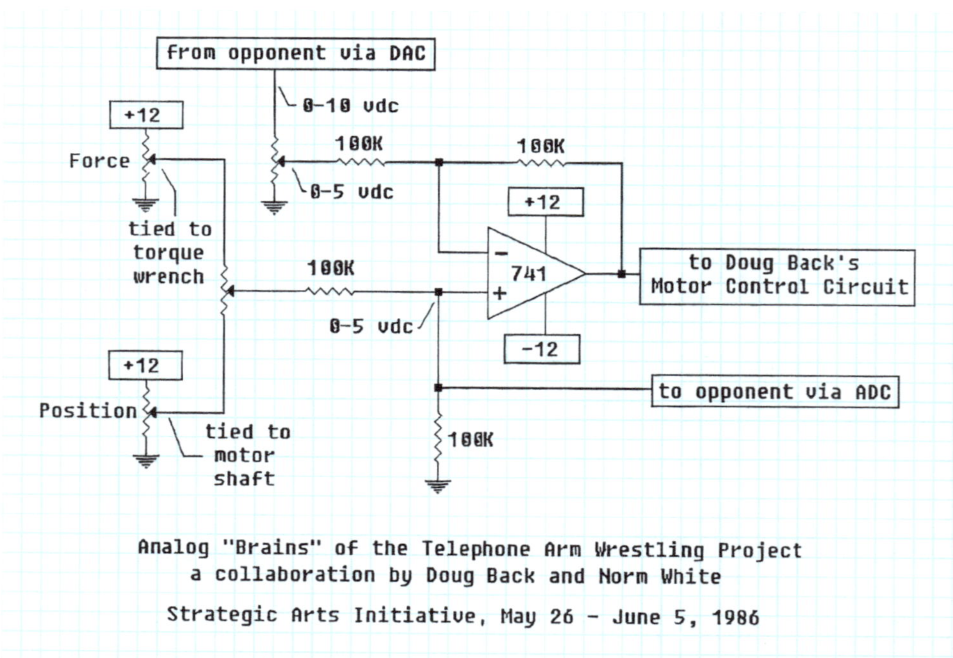

Figure 1.2.4

Electronic schematic diagram of the *Telephonic Arm-Wrestling* system by Doug Back and Norm White. Courtesy Norm White.

attached to a DC motor from a daisy wheel printer. A rotating potentiometer was affixed to the wrench itself to determine the angle of the arm, and with these two values—the force and the position—the trajectory of the arm could be deduced. Finally, there was a 1200 baud modem on each end of the system, which transmitted information to the other arm-wrestling unit through audio on an analog telephone line (figures 1.2.5 and 1.2.6). Back was in charge of constructing the mechanical components of the system, Norm White was in charge of electronics, and they had mutual friend Caroline Langill weld together a support structure for the system. The technician Ian McGuigan also assisted with the project.

As for many media art projects, the first exhibition of *Telephonic Arm-Wrestling* was a technical failure. The two arm-wrestling units did not properly communicate when connected between Canada and Italy. A poor-quality

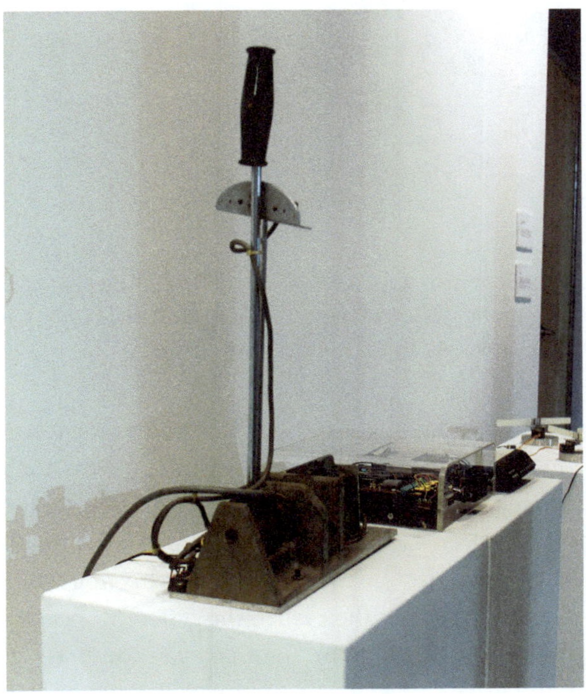

Figure 1.2.5
Parts of the *Telephonic Arm-Wrestling* system on display in the "Influenc(Ed.) Machines" exhibition at the Ontario College of Art and Design University in 2016. Courtesy Treva Legassie, Ontario College of Art and Design University.

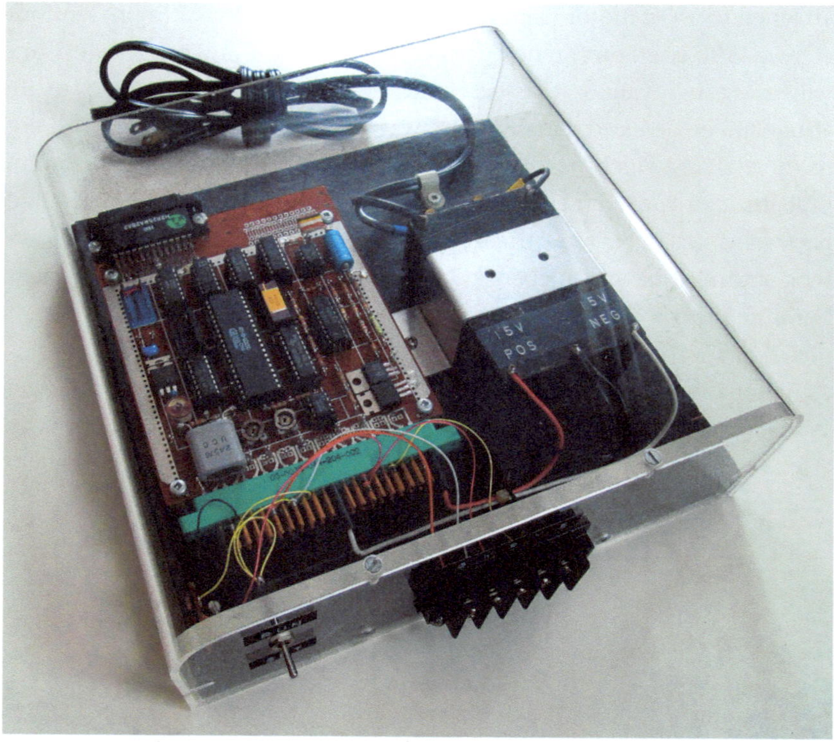

Figure 1.2.6
Detail of the control system for *Telephonic Arm-Wrestling* by Doug Back and Norm White, on display in the "Influenc(Ed.) Machines" exhibition at the Ontario College of Art and Design University in 2016. Courtesy Norm White.

analog telephone connection was the culprit: one side could hardly hear the other when making standard phone calls, which proved too difficult for the modems to reliably operate. After the device's initial failure in connecting participants between Canada and Italy, White claims that Back wanted to abandon the project and was "dragged out of his house" for an exhibition a week later in Paris.[16] To their relief, the system worked successfully at the second show. In the exhibition, the Canadian Cultural Centre in Paris was linked to the Artculture Resource Centre in Toronto, with the live connection active between the two arm-wrestling units for fifteen to twenty minutes.[17] However, due to a half-second latency imposed by a slow transatlantic phone network, the project was not robustly functional as an arm-wrestling device.

In summary, the project concept started out as a traditional test of physical strength, but due to technical limitations it resulted in a machine that was more about sensing presence than about arm wrestling. For example, the relatively high-resolution motion data enabled some users to faintly feel the heartbeat pulse of the person on the other side of the ocean, which resulted in an uncanny human transcontinental presence.[18] Users could also easily beat their opponents by quickly moving their arms, because the split second it took for arm position data to transmit across the Atlantic was enough time for the opponent win before the user could react. The 24-volt DC motors could also be overpowered by most individuals, since they were not as strong as a human arm. In spite of failing as a strength contest, *Telephonic Arm-Wrestling* became one of the first implemented telerobotic pieces in the domain of art.[19]

Jugaad, Function, and Behavior as Artwork

Back and White's project has more in common with an engineering prototype or folk invention than most artworks, because the system is primarily functional, not aesthetic. Norm White summarizes this approach in a four-point statement on his art practice that is worth noting:

1. Art should concern itself as much with behavior as it does with appearance.
2. Some of the best art happens when behavior and appearance are completely at odds with each other.
3. Economy of means is a critical part of aesthetics.
4. Art functions best, and is most needed, outside of galleries and museums.[20]

In this summary of his artistic tenets, White positions function-oriented behavior as a counterpoint to appearance-oriented aesthetics. In many ways, *Telephonic Arm-Wrestling* is at odds with scholarship in the field of art that concerns itself exclusively with visual representation, viewers, and audiences. Instead, using the functioning technology as an experimental user experience (UX) *is* the work. This idea follows a Fluxus-like focus on activity, which has resulted in a history of mechanical and robotic art devices whose core attribute is the attempt to embody a behavior through kinetics.

One way to conceive of this shift away from aesthetics toward functional behaviors is to understand the rise of cybernetics and systems theory in the mid-twentieth century. Art critic Jack Burnham saw this shift as a "transition from an *object-oriented* to a *systems-oriented* culture. Here change emanates, not from *things*, but from the *way things are done*."[21] As a background context for *Telephonic Arm-Wrestling*'s emphasis on function, the most notable early work in this genre is Nicolas Schöffer's *CYSP 1 (Cybernetic Spatiodynamic Sculpture)* from 1956. Schöffer's work was a tall kinetic sculpture with spinning and rotating colored panels that sat on top of a wheeled base that enabled the entire system to move. *CYSP 1* was programmed with several different behaviors: it became excited when it sensed the color blue, when it was in silence, or when it was in intense light. Conversely, it became calm when it sensed red, when it was in a loud environment, or when it was in the dark.[22] This idea that an artwork could *interactively behave* in a specific way was completely novel, and Back and White's project follows a similar mindset: at its core it was absolutely nothing to look at but something to experience as a functioning process and system. If looked at through the lens of contemporary culture, one might term these early works as the origins of interaction design, human–computer interaction, and experience design.[23]

Returning to Norm White's third point about his art practice, he speaks to an economy of means as a core aesthetic. According to White, his work strives to do more with less, is dedicated to functionality, and is focused outside of the museum and gallery space. One versatile term that refers to an appreciation of frugality, functionality, and everyday environments is the previously discussed concept of jugaad, improvised technical solutions. Jugaad is useful in gaining a deeper understanding of both *Telephonic Arm-Wrestling* and the DIY mindset in general.

Killed by Jugaad: The Dangers of Romanticizing Frugality

There is danger in claiming that a jugaad-oriented approach is beneficial in all situations. For example, consider the improvised vehicles in northern India that are built from wooden planks, assorted car parts, and diesel engines originally designed for agricultural irrigation pumps (figure 1.2.7). These vehicles are low cost and made from repurposed materials, but often cause injuries because of poor braking. In a 2013 analysis of road traffic accidents, Australian researcher Thomas Birtchnell estimated that approximately 14 percent of all casualties in the northern Indian city of Aligarh were due to makeshift jugaad motor vehicles.[24] If safety is a concern, a jugaad approach is likely better suited to the development of artwork and

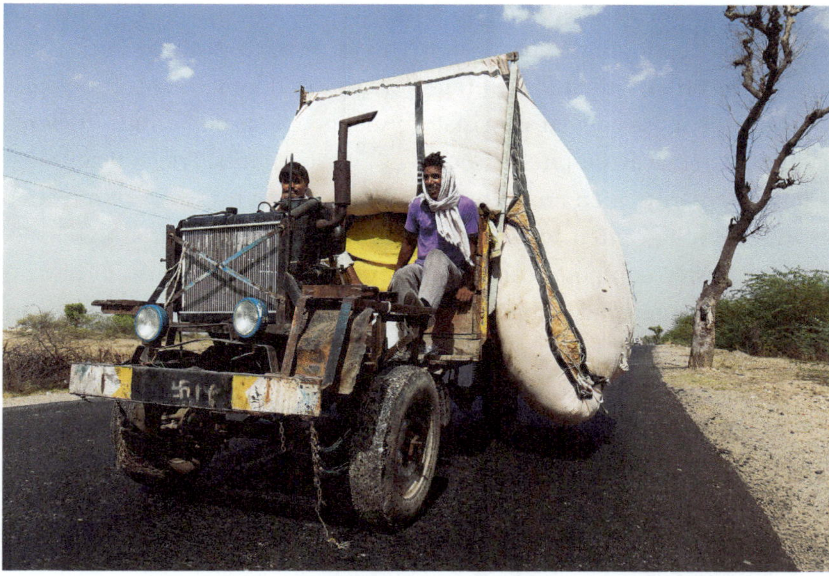

Figure 1.2.7
An example of a jugaad truck made from an assortment of various parts in Madhya Pradesh, India. Unofficially built, homemade DIY-style vehicles are a popular means of transportation in rural India. Photo Studio Octavio / Alamy.

experimental design work than the development of more life-and-death items like vehicles or passenger elevators.

For this reason, Indian academic Rishikesha Krishnan condemns the romanticizing of jugaad, arguing that India must establish itself as a global innovator by moving *away* from jugaad and toward a more systematic and scalable ecosystem of innovation.[25] Similarly, artists like Marc Dusseiller clearly warn against the romanticization of jugaad: "When we think of people doing jugaad we think people are poor and happy but most people are just poor. Jugaad are rare examples that come out of misery. In India it's related to the caste system in the sense that jugaad is done by people who manage to do something despite their lack of education due to caste discrimination. Again, these are a few solutions out of a million impoverished lives. So we should not romanticize it too much."[26]

However, in the context of making artwork, we might recognize as a benefit what Krishnan and Dusseiller see as a deficit. A core takeaway for media art, human–computer interaction research, and experimental design is that jugaad provides an alternative to concepts like *innovation* as the primary model for developing technologies. This is worth repeating: jugaad may not be the official method for professionally developing new technologies, but it is the most popular way that normal people make things—they just start hacking away despite not having many resources. Frugal technological improvisation is an unavoidable part of building things in an imperfect world. Although often invisible, approaches similar to jugaad live inside cut and pasted computer code, as the hot glue holding a custom piece of clothing together, and behind the glossy veneer of a piece of electronic art. The aim here is not to praise jugaad-type methods as the cure-all approach to building things, but rather to clearly acknowledge that most of the time when humans are trying to build or fix something, they normally lack the proper tools, resources, time, and money.

Jugaad Conclusions: A Critical Design and New Media Counterpoint

Jugaad is particularly productive as a counterpoint to ideas like speculative design or new media, which tend to highlight professionally refined physical objects.[27] Exhibitions of glossy technologies may be at home in a museum, showroom, or gallery, but they bear little resemblance to the DIY history of electronic art. In contrast to new media, *Telephonic Arm-Wrestling* is jugaad because it has no embellishments, extra lights, or features, and its frugal materials only strive for mechanical function. In contrast to critical design, it bears little visual resemblance to the "consumer product" aesthetic of the bulk of critical design work. Jugaad-like practices are almost completely devoid of aesthetic embellishments because the goal is to make them work regardless of what they look like or what others might think of them. Jugaad comes from a place of limitations and brutal functionality. For this reason, it fills in the deficiency art theory and technology design have for robust terms to describe constraints.

Although jugaad should not be embraced as a catchall category for all technical or design work, it is considerably useful in describing the messiness of how electronic art projects are built. Despite being constructed with virtually no budget, *Telephonic Arm-Wrestling* successfully transmitted physical movement between two transatlantic participants. Further, it had a very high level of resolution—White claimed that he could remotely feel Back's blood pulsing in his arm, which produced an uncanny sense of presence connecting the participants. As a result, this project stands as one of the earliest in the field of robotic telepresence, and serves as an outstandingly useful example of constrained DIY and an illustration of the creative technical potential of working with and around limitations.

To accept jugaad is to accept imperfection and to acknowledge that few individuals have the resources to build professional-looking "new media" devices. It also provides a concept for understanding how a vast majority of the world's population creates things through hacking.[28] Jugaad work is not just post-optimal (Dunne), it is nonoptimal—it never had the opportunity to be optimal and does not strive to be perfect. In this way, jugaad pushes back against design's obsession with refined perfection.[29]

1.3 Frugality and the *Toaster Project*: Technical Disorientation, Device Paradigms, and Highlowness

Thomas Thwaites's *Toaster Project* (2009) is an exploration of an individual trying to build a contemporary toaster by oneself (figure 1.3.1). The designer built a simple, electric, domestic appliance completely from scratch from within the UK—literally digging the iron ore from the ground and trying to create plastic, for example. In reviewing the original motivation for the project, Thwaites states that he "wanted to get under the skin of the slick-looking objects that surround us, but don't really come from anywhere."[1] In the process, the *Toaster Project* provides a stark example of the generative creativity of constraint-by-choice.

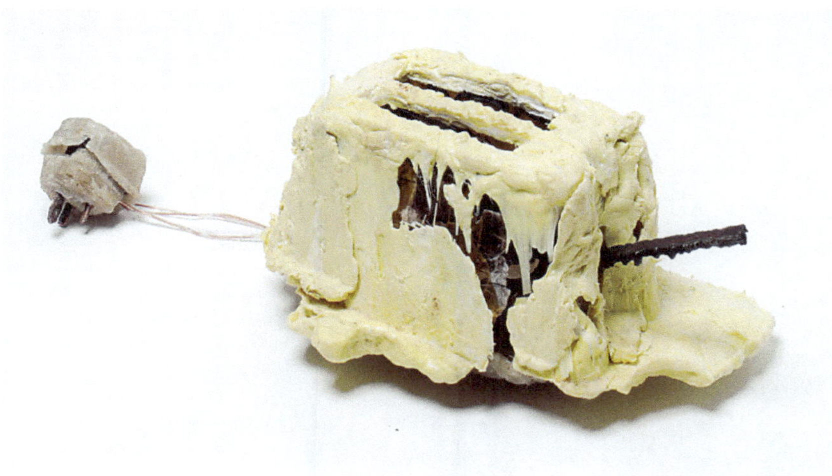

Figure 1.3.1
The finished toaster produced as a part of the *Toaster Project* by Thomas Thwaites. Photo Daniel Alexander. Courtesy Thomas Thwaites.

Rules-Based Constraint: The *Toaster Project*

Thwaites started the project in September 2008 by purchasing the cheapest toaster he could find at a local shop: an "Argos Value Range 2-Slice White Toaster" for £3.94. He carefully disassembled the toaster into 157 main parts, but after inspecting these components further, he broke the main parts down into 404 individual pieces (figure 1.3.2).[2] By grouping the parts into their base materials, Thwaites felt that he could roughly reproduce the toaster by sourcing five key materials: (1) steel for the toaster subframe and pop-up mechanism; (2) nickel for the heating element; (3) mica—a natural electrical and thermal insulator—for the heating element to wrap around; (4) copper for the electrical wires and plug; and (5) plastic for the exterior case of the toaster. With a focus on sourcing steel, nickel, mica, copper, and plastic, Thwaites set off on building a toaster from scratch with three core guiding principles:

Rule 1. My toaster must be like the ones they sell in the shops.

Rule 2. I must make all of the parts of my toaster starting from scratch.

Rule 3. I will make my toaster on a domestic scale.[3]

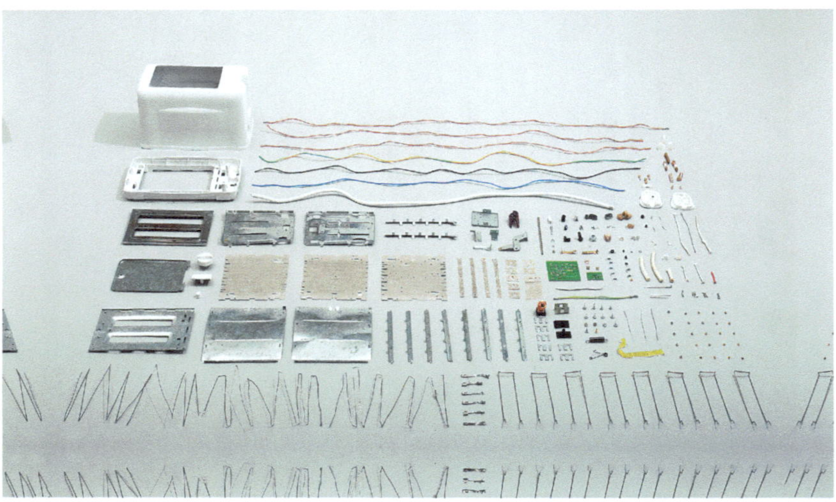

Figure 1.3.2

Thomas Thwaites's meticulously disassembled "Argos Value Range 2-Slice White Toaster." Photo Daniel Alexander. Courtesy Thomas Thwaites.

In other words, he forbade flying and restricted himself to only traveling overland. He also restricted himself to using only domestic-grade tools, which ruled out tools like professional steel refining equipment. Thwaites set strict DIY parameters on his project in an attempt to explore the layers of infrastructure within a humble, low-cost, household consumer product.

Thwaites spent the next nine months tracking down these five basic materials within the UK and worked to try to refine and model them into a functional toaster. He dug raw iron out of a defunct mine in England and smelted it down using a microwave oven. Mica for the electrically insulating component of the toaster's heating core was sourced by chipping away at a hillside in Scotland. The *Toaster Project* is unique from other projects discussed in this manuscript so far in that its development was purposefully constrained by a clear set of rules. This regulation-oriented method of developing a project—where self-made guidelines are determined by the creator from the start and blueprint the entire development process—is a common technique for artistic development. Self-imposed rules can reflect many different forms of limitation, including restricted materials (for example, making a chandelier out of only paperclips), restricted processes (making a drawing by holding a pencil with your feet), or restricted time and location. Limitations naturally exist during the process of building any objects, but what Thwaites illustrates more resembles a game of self-imposed rules.

Reductio ad Absurdum

Anthony Dunne and Fiona Raby refer to Thwaites's toaster as "*reductio ad absurdum,* a type of logical argument in which one assumes a claim for the sake of argument and derives an absurd or ridiculous outcome by taking it to its extreme."[4] Literally "reduction to the absurd" in Latin—*reductio ad absurdum* is an argumentative method for showing that something is false by illustrating that its logical consequence is monstrous or contradictory.

The *Toaster Project* reduces many different concepts into something like a DIY monstrosity (figure 1.3.3). One idea that is reduced is that artisanal products are better than factory-built ones. While Thwaites's toaster arguably has more character than the original "Argos Value Range 2-Slice White Toaster," it certainly has none of the toasting functionality: Thwaites's toaster does not electrically function. Yes, it is locally and artisanally made, but as an operational device it is an abject failure.

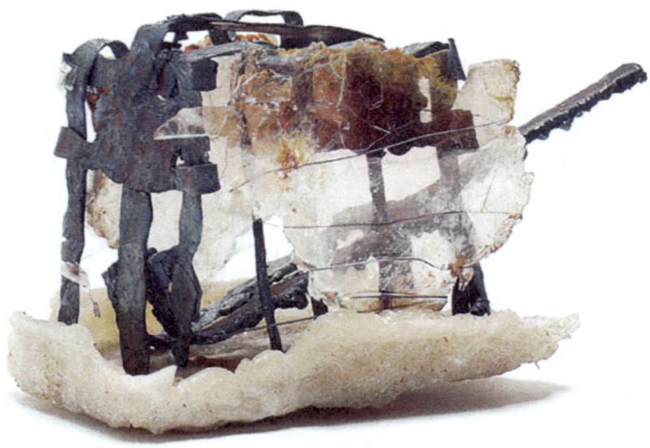

Figure 1.3.3
The finished toaster with its cover removed, by Thomas Thwaites. Photo Daniel Alexander. Courtesy Thomas Thwaites.

Another reduced idea is the assumption that humans are masters of contemporary technology—and that they can comfortably function independently of mass-manufactured consumer products. Thwaites's toaster highlights the complexities of simple conveniences like toasting a piece of bread and the extent to which we are often distanced from doing something completely on our own, especially the construction of a consumer-grade product. Thwaites's toaster can be thought of as a response to what Sandra Alfoldy terms "craftwashing"—the corporate and mass uptake of terms like "handmade," "small batch," "artisanal," "bespoke," and "craft," and the public's almost unquestioning acceptance of these terms.[5] Thwaites gives us an "artisanal toaster" that is all effort and no function, and in the process he raises the question of whether or not we want everything to actually be handmade. Do we want devices that resemble an Iron Age (or *Gilligan's Island*) attempt at making a contemporary electronic device?

DIY Practice and Borgmann's Device Paradigm

A useful framework for thinking about products within the context of Thwaites's Toaster includes Albert Borgmann's concepts of blackboxing and device paradigms:

> Technological information . . . rests on a substratum of machinery that is becoming concealed from the understanding of those who operate on its surface. The blackboxing that is the consequence of progress in information technology

> encloses ever larger spaces of hardware and software. It is an unavoidable development. The larger black boxes support more powerful tools.[6]

Increasingly concealed, the inner workings of consumer electronics and information technologies are a result of the development of newer generations of technologies. Devices are built out of existing technologies, and in the process, the components fade from being contemplated objects into the background of infrastructure. In other words, as technologies become easier to use, they often become harder to figure out.

Technology, as it is commonly understood, "promises to bring the forces of nature and culture under control, to liberate us from misery and toil, and to enrich our lives." Technology often promises liberty and prosperity as long as it is quick to access, available everywhere, safe to use, and easy to understand. "In the common view, technological progress is seen as a more or less gradual and straightforward succession of lesser by better implements."[7] As technology develops, the mechanism of a device tends to become concealed.[8]

During the development of advanced technologies, the means and materials of the device get separated from the end use of the object—in most consumer products they are distinct and separate from each other, with the function of the product taking priority. Borgmann presents an example of a watch: it can be mechanical, electronic, or exist as a mobile phone app, with highly variable mechanisms behind each—but Borgmann sees the functional device paradigm as driving the device, its "watchness." The "softwareness" or "brassness" is less of an issue for the end user than its ability to keep time. Devices can be radically morphed without killing the category of the thing—"of radical variability of means and relative stability of ends."[9] Borgmann sees this as the "device paradigm," where most relationships to technological objects consider them devices to accomplish a specific task, as opposed to understanding objects as actual technical things with specifically engineered biases or as objects sourced from particular materials. Similarly, Andrew Feenberg describes this shift into blackboxed technologies as being where products "no longer realize potentialities within a world but are stripped bare of qualities, of their very thinghood, to take their place in a technological system."[10] Thwaites takes the object out of its technological system and reverts it back into bare "thinghood," and the resulting process reveals the complex web of social and material links that are bypassed when the toaster is only used to make toast.

Borgmann uses the term "focal things and practices" to denote objects that, instead of just being viewed as commodities or devices, are made central, clear, and articulate.[11] A key component of focusing on things is the development of skepticism toward advanced technologies: "If we recognize the central vacuity of advanced technology, that emptiness can become the opening for focal things."[12] A focused engagement, in other words, is the flip side of only thinking of technology as blackboxed devices.

A DIY-created object like Thwaites's toaster operates as a focal thing that complicates the device paradigm of affordable domestic conveniences: toasters are profoundly complex when looked at in detail, even though they are one of the simplest appliances. The convenience of easy-to-use consumer products come at a cost: they divorce us from the economics, tradition, interpersonal relations, social fabric, and rituals associated with material culture. Blackboxed products often give little indication of the commodities that built them and the complex web related to how they were produced. Thwaites completely breaks these boxes open and splays them out in front of us—the impossibility of the opened and dissected black boxes of contemporary consumer products is the heart of Thwaites's project. Borgmann sees technology as powerfully separating us from the means of how something is produced, of splitting means and end, which Thwaites makes a valiant attempt to rejoin. Projects that focus on unblackboxing or unpacking closed systems work to convert commodities back into objects and things and to reconnect them with new economics, traditions, and interpersonal relations. Although the artists and designers in this book are motivated by radically different cultural factors and economies than would be someone like a backyard auto mechanic in a developing region, the two groups are similar in the formation of their own nonofficial networks of craft that delve inside proprietary systems, modify them, and reassemble them back into nonstandard but more useful devices than the original.

DIY work done under material, financial, or rules-based constraints shows how technologies and objects are often reused and hacked, and in the process, ignore or invert their anticipated uses as commercial products or pieces of industrial design. As a result, they challenge the conceptual models embodied in the design of the original product or system.[13] Kludged or cobbled-together devices are like a folksy cul-de-sac standing apart from the flow of progress, and as such, they have the opportunity to open up everyday objects into focal things.

Disorienting Technologies and the Burden of Doing Things Yourself

The *Toaster Project* does not simply encourage us to consider where and how everyday devices are made. The *Toaster Project* also warns us about the dark side of digging down into the underlayer of "thingness"—like Thwaites, we might end up spending £1187.54 and traveling 1900 miles over nine months only to end up with a god-awful mess that does not work.[14]

The extravagance of the *Toaster Project* is a larger scale version of do-it-yourself endeavors in general: they are often horribly inefficient uses of time, energy, money, and resources. This is familiar territory to those who have attempted do-it-yourself home repair or to those who have tried baking an ornate cake for the first time. DIY practices typically start with the promise of saving money and building a thing exactly the way you want it, but after the complexity of the task becomes disorienting, it usually ends up costing more and working less well than a professionally made equivalent.

Borgmann's concept of DIY disorientation is also useful in thinking through this component of the *Toaster Project*. Borgmann sees DIY practice as potentially exhausting and disorienting, especially when trying to engage deeply with technology and also trying to be self-sufficient. Building, rebuilding, refining, and maintaining objects "on which nothing is ever enacted" can serve as a distraction or a peripheral engagement that drowns out or kills the enjoyment, fun, and celebration of life.[15] In the case of Thwaites, his project had an endpoint that resulted in a graduate degree from the Royal College of Art, being interviewed by Stephen Colbert, and a book from Princeton Architectural Press. If only measured in terms of the effectiveness of his toaster—and not the epic journey of his effort—Thwaites's DIY journey might not have been as rewarding.

DIY with a focus on radical self-sufficiency runs the risk of strangling out our relationships and our ability to engage in the political and cultural world. Instead, Borgmann sees that a balanced acceptance and adoption of technology, or an intelligent limitation of how we adopt technologies, is preferable. Borgmann sees that "one should gratefully accept the disburdenment from daily and time-consuming chores and allow celebration and world citizenship to prosper in the time that has been gained."[16]

The freedom of self-sufficiency is a consistent trope for DIY endeavors—and Thwaites's toaster provides an elegant visual reminder of what "burdened" self-sufficiency looks like. The toaster is a reductio ad absurdum of radical self-sufficiency, and it reminds us that doing everything yourself does not necessarily result in an increased quality of life.

Extreme Self-Sufficiency and Open Source Overload: The Global Village Construction Set

The do-it-yourself allure is the thought that because it provides freedom in certain situations, it will provide more freedom when applied to everything. An example of this can be seen in Marcin Jakubowski's *Global Village Construction Set*—an endeavor to design and build an entire open-source ecosystem of DIY machines such as tractors, ovens, and electronic circuit makers—with the goal that "anyone can build and maintain machines at a fraction of what it costs today."[17] The ambitious plan for the *Global Village Construction Set* is to take the DIY approach used to build open-source 3D printers and apply it as a blueprint for civilization—or at least the fifty most important machines the organization thinks are needed for modern life to exist (figure 1.3.4). Jakubowski's project to build "powerful, self-replicating production tools—that can decentralize production—to build modern prosperity in local economies" is both alluring and motivated.[18] Borgmann and Thwaites might see Jakubowski's entire enterprise as completely misguided, or at least headed down the path of promising convenience similar to a toaster but actually being next-to-impossible to DIY all of the physical tool infrastructure a society needs on one's own. Physical objects have such complex supply chains in comparison to software development that some open source researchers like Jérémy Bonvoisin suggest that using the term "open source hardware" is problematic—it is often unclear what the source of open source hardware actually is.[19]

A common mistake in design work is thinking that others have the same abilities and resources as you, which can be described as *Déformation professionnelle,* or a cognitive distortion based on your skillset or profession.[20] When Jakubowski states that "anyone can build and maintain machines at a fraction of what it costs today," what he means in practice is that "I hope I can build and maintain machines at a fraction of what it costs today."[21] Jakubowski's 3D modeling skills, education, and extensive mechanical

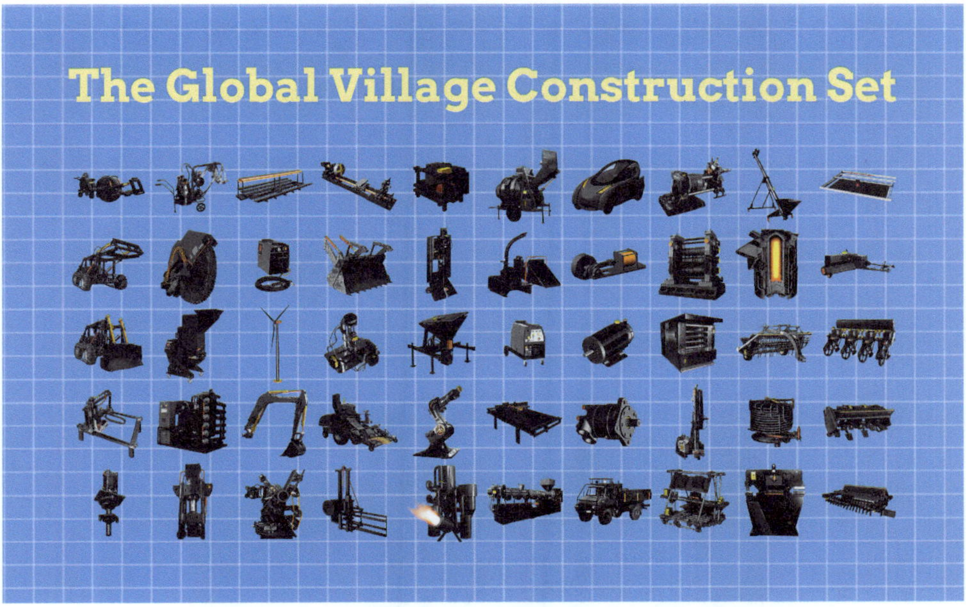

Figure 1.3.4
The *Global Village Construction Set* is a project initiated by Marcin Jakubowski in 2003 that envisions itself as a "civilization starter kit," in other words, a set of the fifty basic machines it takes for modern life to exist—everything from a tractor to an oven to a circuit maker. Courtesy Marcin Jakubowkski, Creative Commons Attribution-ShareAlike 4.0 International (CC BY-SA 4.0) License.

fabrication experience, plus his space, money, time and help from volunteers do not get transferred when his CAD files or source code is shared publicly. When an author's experience and resources are stripped away from a DIY initiative, the project dramatically shifts. For example, what would happen if Thomas Thwaites attempted to build his own tractor, oven, or electronic circuit maker?

The *Toaster Project* offers a considerably different picture of do-it-yourselfness than a project like the *Global Village Construction Set* (figure 1.3.5). Thwaites's portrait of DIY starts from a point of attempting to be completely independent, to vigorously try to do everything yourself. Although Thwaites was arguably quite well equipped, his strict rules of self-sourcing materials and only using domestic-grade equipment resulted in a completely relatable result: a DIY "fail." Do-it-yourself processes often end up

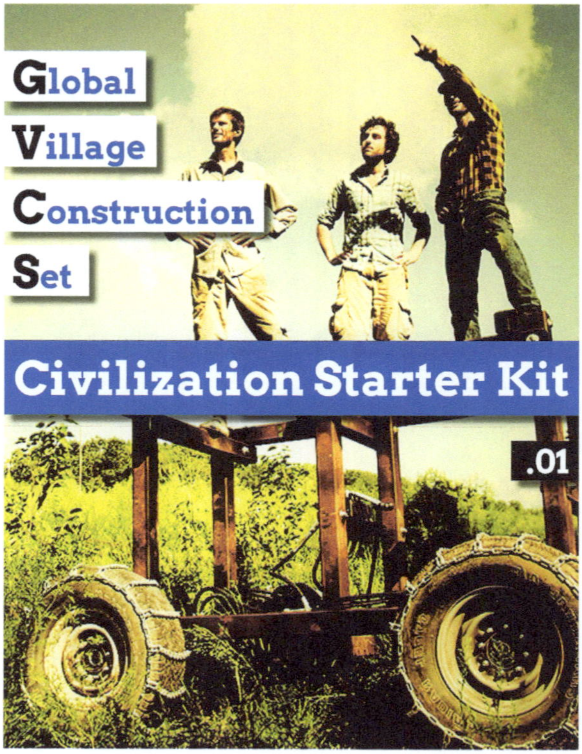

Figure 1.3.5
Cover page of the *Global Village Construction Set Civilization Starter Kit .01*. Courtesy Marcin Jakubowski, Creative Commons Attribution-ShareAlike 4.0 International (CC BY-SA 4.0) License.

in a mess when we dig deeply into the making process. When attempting to build something on our own from scratch, the complexity of the system explodes. The *Toaster Project* reminds us that we often only skim the surface of making something when we build it, and that we often require vast experience, money, and infrastructure if we hope to have the thing actually function. Thwaites's *Toaster* reminds us of the inefficiencies of doing everything yourself, and the risk of getting technically overwhelmed in the process and making nothing of much use beyond an admirable monument to human effort.

However, the danger of projects like Jackowski's is that they can easily promise freedom at the expense of neglecting the complexity of technical infrastructure and support systems. Given the choice between Thwaites's *Toaster* and Jackowski's *Construction Kit*, the toaster is a clearer rendering of how technological systems tend to work—I would argue that the default human behavior is to initially fail at most technical tasks, with a mess made in the process. Humanity is more about the beauty of failure than a self-supportive open source global construction kit. The open source global construction kit already in place is the ordinary jugaad, DIY and ad hocism all around us.

Conclusion: What We Can Learn from an Artisanal Toaster

Thomas Thwaites's *Toaster* reminds us of many things. First, it gives us an image of what "true" DIY looks like, or at least an attempt to build something completely from scratch. The lesson is that "completely from scratch" is virtually impossible—DIY is never completely independent and always sits on top of mass-market industrialized convenience. However, reaching down into the lower levels of materials or process can, as in the case of Thwaites, break us out of the device paradigm of just thinking of the objects around us as devices to accomplish something else. This is useful in revealing and dissecting the infrastructures around us that are commonly invisible, but also comes at the cost of significantly disorienting us. In other words, DIY methods can be a useful way to reset our assumptions about infrastructures, but this reset can overwhelm us. This "highlowness"—of reaching down and impossibly transforming basic materials into something complex—is a noteworthy dynamic in DIY practice that can be done with slapstick chindogu-like humor or the seriousness of a mystical alchemist. In the end, sometimes DIY work is more about human effort than human accomplishment.

Theme 2 Exploring Technologies

Exploration and DIY Electronics: A Thematic Overview

All contemporary technologies are a bit of a "black box." In other words, the interior of how technologies work is often a mystery—like a sealed flight data recorder with a hidden and inaccessible interior (figure 2.1). For example, when one uses a device like a telephone, the complex technical workings of the electronics, networks, and infrastructure are completely invisible: you simply enter the phone number and the call is completed. Once developed and deployed widely, technical components are understood by users as objects that serve a particular function. An electronic toy makes a sound when a button is pressed, a television displays a channel, and a printer outputs a document when requested.

Within the disciplines of engineering and computer science, a *black box system* refers to something that only exists as an input and an output; blackboxed systems have internal workings that are concealed or opaque. In computer science terminology, for example, "blackbox testing" refers to testing that is done by end users who have no knowledge of the computer source code, while "whitebox testing" refers to when testers can see the source code: "The white box tester (most often the developer of the code) knows what the code looks like and writes test cases by executing methods with certain parameters."[1]

Blackboxing, or the development of technological objects to the point where they are simply used and not understood as technical objects, is often a way to manage the development of complex technologies. A computer system, for example, is incomprehensible to any one person trying to understand how millions of transistors and circuits interact with software. Black boxes are also the building blocks from which new technologies and infrastructures are

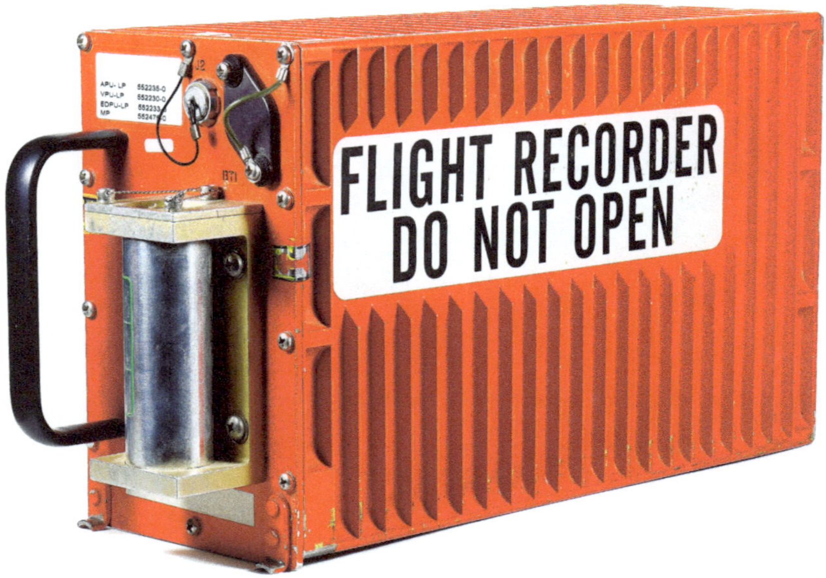

Figure 2.1
An image of a flight data recorder, often referred to as a "black box." Data recorders are an obvious example of a system where the interior functionality of the device is sealed off and not accessible to a user. A flight data recorder is engineered to be tamperproof and to operate autonomously. Photo Björn Wylezich / Alamy.

constructed. In *Pandora's Hope*, philosopher Bruno Latour defines blackboxing as a necessary byproduct of a technological system. "When a machine runs efficiently . . . one need focus only on its inputs and outputs and not on its internal complexity," he writes. "Thus, paradoxically, the more science and technology succeed, the more opaque and obscure they become."[2]

The Bringing Together of Punctualization and the Pulling Apart of Depunctualization

If the metaphor of a network is used to describe this process, one can say that blackboxing essentially takes a network of things and simplifies it down into a single node. The complexity of a network of things is consolidated into a singular point that is simpler to use, with the tradeoff of it being more difficult to understand. "Punctualization"—which might be thought of as "bringing things together into a point"—is a useful term to describe this process. Punctualization can be defined as the process of a complex technology becoming an easy-to-use component that can be easily used as a building block to make more complex things. A popular example is the computer system: computer systems were initially developed as large physical things but have become less expensive and easier to use and are embedded in many devices today. We can say that computerization has been punctualized in different ways: for example, building an Internet of Things device is considerably easier when one can purchase a computer chip like the Espressif ESP32 for about $5 that is a plug-and-play computing platform with Wi-Fi and Bluetooth.[3]

The term punctualization is used within sociology by Michel Callon and John Law to describe how networks of things converge into a single entity: "The process of punctualization thus converts an entire network into a single point or node in another network."[4] This simplification offers many advantages: "Punctualized resources offer a way of drawing quickly on the networks of the social without having to deal with endless complexity."[5]

The core takeaway for DIY practice is that although some artists and makers are focused on bringing technologies and networks together into punctualized systems that can serve as building blocks for others, the more important cultural and technological role within DIY electronic practice is in breaking black boxes apart. The opaqueness of technology is a core theme that DIY electronic practice, as I have described in this book, pushes against. Artists and designers build projects that question the closed functionality of technological systems. As a result, many contemporary artists working with electronics can be thought of as unblackboxing or depunctualizing technologies: cracking open systems and bending, playing with, and exploring rewired interiors. Far from being a purely technological exercise, I argue that unblackboxing serves vital social and political functions through DIY practices.

An Unblackboxed Television: *Magnet TV*

An early piece by Nam June Paik (1932–2006) titled *Magnet TV* (1965) demonstrates how unblackboxing and DIY-style practices operate within the field of contemporary art. *Magnet TV* was first publicly shown in Paik's first solo exhibition, titled "Electronic Art," at Galeria Bonino in New York City in 1965. The piece has a large, thick, horseshoe-shaped iron magnet on top of a large black-and-white TV set (figure 2.2). The side of the television cabinet has a large hole in it with rough, gnawed edges and a piece of clear plexiglass forming a window to view the set's tubes and electronic components. Viewers were invited to "modify the television's output into swerving abstract lines through the movement of a magnet over the TV"[6]—manually moving the magnet to twist the screen image into an elegant Möbius strip-like ribboned shape (figure 2.3). The immediate effect of the magnet on the path of the electrons, alongside the opening in the box of the television cabinet, reveals its inner technological workings.

Magnet TV highlights how layers of technology can be unblackboxed through simple tools and DIY techniques. In this case, a magnet and a hole on the side of the set unravels a piece of standard consumer electronics into something much more complex: the punctualized black box is opened up. This process of depunctualizing strips out the standard use of the device and reverts it back into an electromechanical thing. *Magnet TV* seeks interactivity and has sparse elegance: "What made *Magnet TV* so successful with critics was not only that they could see the principle behind

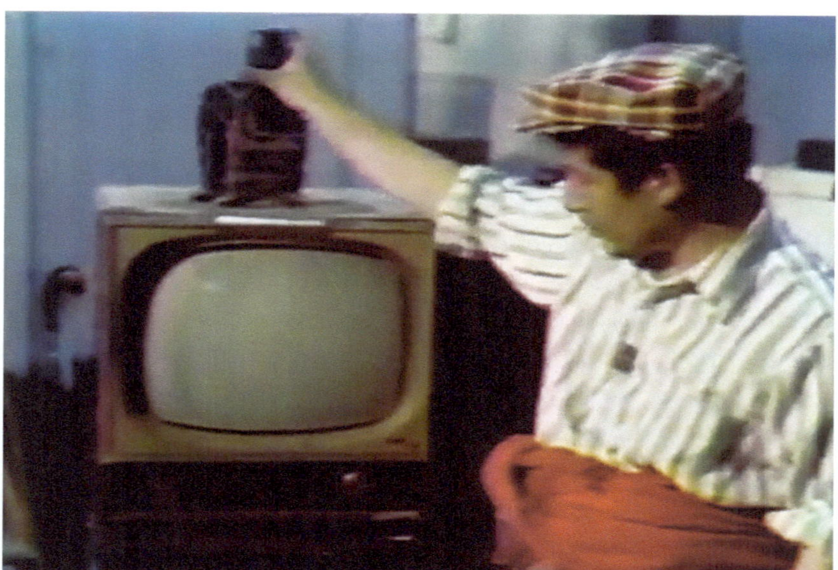

Figure 2.2
Magnet TV manipulated by Nam June Paik. Courtesy © Nam June Paik Estate.

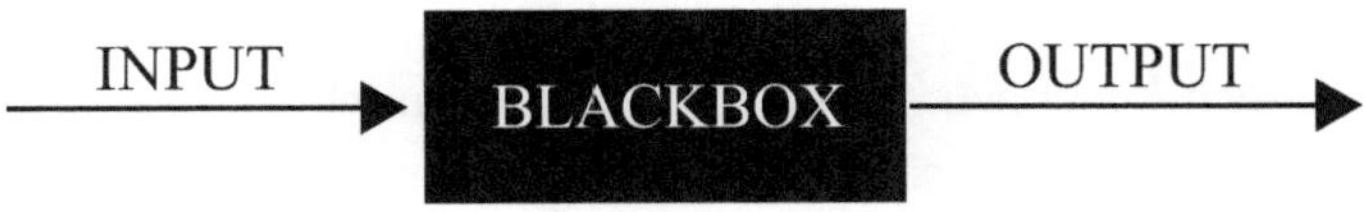

Figure 2.3
A blackboxed system processes input into output without the user's knowledge of the interior functionality of the object. Image Garnet Hertz.

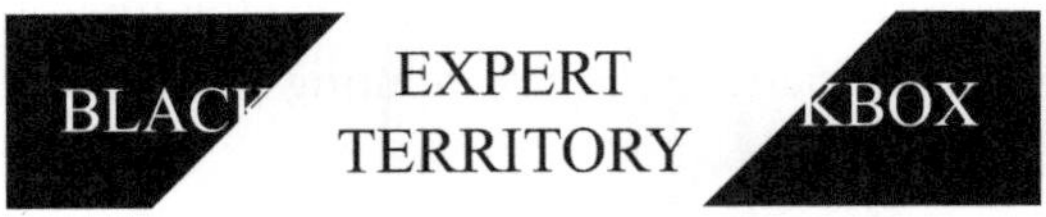

Figure 2.4
The interior of a blackboxed system is expert territory and tends not to be user serviceable. In contrast, devices that are "hackable" have an interior that is not exclusively expert territory and has some user-serviceable parts inside. Image Garnet Hertz.

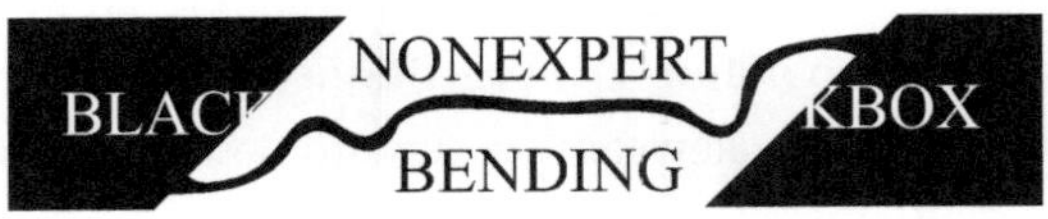

Figure 2.5
Despite being expert territory, portions of the non-user-serviceable interior of blackboxed systems can be manipulated and bent by nonexperts. This exploratory process is a proposed key theme in the practice of circuit bending and other DIY electronic practices. Image Garnet Hertz.

it at first glance, but also that a simple artistic intervention which anyone could carry out had created pictures of such stunning aesthetic charm."[7] *Magnet TV* reminds us that it does not necessarily take considerable technical skill to unblackbox technology.

Visualizing the Black Box

If thought of as a system and visualized as a flowchart, a black box typically has some form of input, a big opaque box in the middle, and some form of output that comes out of the box (figure 2.4). In the case of a classic television set, the user input is in the form of volume and channel knobs and the output is the picture and sound. The interior functioning of the system is hidden from view and not required to be understood.

Magnet TV unblackboxes a standard television in two main ways: first, it literally punches a hole in the side of the wooden television box to visibly expose its electronic parts. Second, the magnet bends the electronic flow within the picture tube and makes the interior of the system more visible by destabilizing it. These two techniques are a few of the various ways that artists and experimental designers open up and explore the interiors of technologies—and this type of unblackboxing is a core theme throughout creative DIY electronic practices. In many ways, the input-output conceptual model is the core foundation of computing.

Exploring Technologies: Ghazala, Burgoyne, and Hewitt

The interiors of black boxes, generally speaking, are expert territory. Despite this, nonexperts can open up and manipulate the systems in various modes, including something as simple as placing a magnet on top of a television (figure 2.2). Stepping through a number of additional examples gives a clearer indication of how DIY practitioners unblackbox technologies in contrasting ways. Three key case studies are fleshed out in the chapters ahead to help clarify the leverage of DIY practices in culture: Reed Ghazala's *Incantor* (1978), Diana Burgoyne's *Wire Figures* (1985), and Darsha Hewitt's *20 Oscillators in 20 Minutes* (2012).

Ghazala's *Incantor* is a children's electronic toy that is modified into a musical instrument by creatively rewiring its internals. This project and others are discussed in chapter 2.1 titled "Exploration and the *Incantor*: Bending Circuits, Depunctualization and Unblackboxing." Next, Burgoyne's *Wire Figures* consists of humanoid sculptures built out of wire and electronic components that emit audio when they are touched by the public. This is explored in chapter 2.2, titled "Exploration and *Wire Figures*: Transparent Technologies, Interactivity, and Radio Shack Cybernetics." Finally, Hewitt's *20 Oscillators* is a live performance-like demo where she constructs an audio circuit in front of a watching audience. This is described and expanded on in chapter 2.3 titled "Exploration and *20 Oscillators in 20 Minutes*: Technological Performance, Hedonization, and the Thrill of Impending Failure." Each of these projects shines light into concealed components of technology. *Incantor* illustrates the beauty of mistakes and turning errors into a feature, Burgoyne's *Wire Figures* builds a nonopaque "transparent box" that humanizes electronics through public interaction and handmade care, and

20 Oscillators in 20 Minutes gives us a performative, step-by-step building of a complex system with humor spilling out of the gaps and errors of trying to do something yourself.

All of these projects invite us to come in and explore the black box of technology. In the process, they illustrate how DIY practitioners crack open and adapt technologies for highly creative uses. Although there are significant limitations to the process of unblackboxing, it reminds us that we do not need to accept technologies and systems at face value. The interiors of technology hold ample opportunities for playful misuse, reconfiguration, and rebuilding. The DIY mindset is, in many ways, an unblackboxed mindset. In these ways, each of these artists reveal hidden forms of knowledge inside of punctualized objects, and in the process they show us the leverage that small-scale and informal technological production has in contemporary society.

2.1 Exploration and the *Incantor*: Bending Circuits, Depunctualization, and Unblackboxing

> Technological information . . . rests on a substratum of machinery that is becoming concealed from the understanding of those who operate on its surface. The blackboxing that is the consequence of progress in information technology encloses ever larger spaces of hardware and software. It is an unavoidable development. The larger black boxes support more powerful tools.
>
> —Albert Borgmann, *Holding on to Reality*

The inner workings of consumer electronics and information technologies are increasingly concealed as a result of the development of newer generations of technologies. Devices are built out of existing technologies, and in the process, the components fade from being contemplated objects into the background of infrastructure.

When mature and deployed widely, technologies are understood by users as objects that serve a particular function: a doorbell makes a sound when pressed; a telephone makes a telephone call; and a photocopier outputs a document without much complex technical negotiation. The inner workings of the technical thing are unknown to the user, with the circuitry of the device acting like a mysterious "black box" that is largely irrelevant to its use. It is only an object with a particular input that results in a specific output; its mechanism is invisible. From a design perspective, the technology is normally designed to render the complex mechanisms invisible and usable as a single punctualized object.1 As a part of this evolutionary process, black boxes are the punctualized building blocks from which new technologies are built.[2]

Bending Black Boxes

A black box, however, is a system that is not intricately understood or accessed, and as a result, obsolete or broken technologies are often unusable. Once the product stops working, the device is often unrepairable and inaccessible by the average consumer. Unlike a household lamp which can be fitted with replacement light bulbs, many consumer electronic devices have no user-serviceable parts, and the object is discarded after it breaks. The depunctualization, or breaking apart of the device into its components, is difficult due to the highly specialized engineering and fabrication processes used in the design of the artifact: electronic technologies are typically engineered to be discarded instead of repaired. However, artists, musicians, and other DIY practitioners frequently take technological obsolescence as a starting point for their work. Obsolete objects are regularly disassembled, rewired, and repurposed for different applications than what they were originally designed, and in the process, the black boxes are "bent" into new forms.

Ghazala and the Origins of Circuit Bending

Artists, musicians, and creative individuals probe and explore obsolete technologies in ways that expand our standard notions around what is technologically obsolete or useful. Old computers, vinyl, cassette tapes, or pieces of retro musical equipment are reused, repaired, and remixed with newer components. Media artists and creative hackers regularly manipulate and customize technologies for the fun of exploring the technological system, often breaking apart and reverse engineering without formal expertise, manuals, or a defined endpoint. One noteworthy figure on this topic is Reed Ghazala, a Cincinnati-based American artist born in the 1950s. He has been pivotal in the formation of the practice of "circuit bending." The DIY technique of circuit bending takes found objects like battery-powered children's toys and creatively repurposes them into spectacularly original musical instruments and homemade audio generators.

Circuit bending, a term coined by Ghazala, is a methodology for modifying inexpensive secondhand circuits that he first explored in Ohio in the late 1960s. The origin of his method came as the result of a random amplifier malfunction that occurred while he was a high school student (figure 2.1.1):

> In my drawer a small battery-powered amplifier's back had fallen off, exposing the circuit. It was shorting-out against something metal causing the circuit to act as an audio oscillator. In fact, the pitch was continuously sweeping upward to a peak, over and over again. . . . I soon modified the amplifier in numerous ways. Placing the circuit within a larger housing, I added rotary switches to the short circuit paths so I could run the new circuits through various resistors, capacitors, diodes, photo cells, and any other electronic component I could find. Potentiometers and push-buttons were added. I discovered places on the circuit that, if touched, would make the circuit howl.[3]

Bending developed as an experimental and unpredictable process to reuse, reappropriate, and customize scrap electronics, placing value on highly

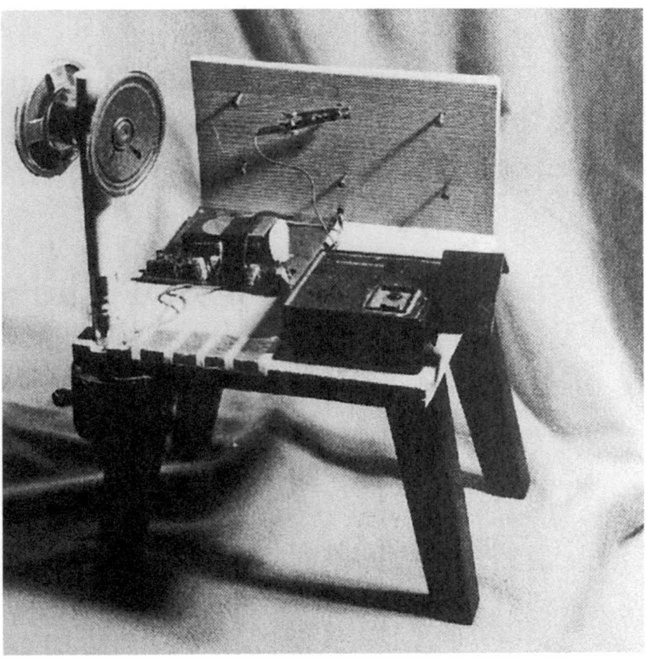

Figure 2.1.1
Ghazala's reconstruction of his first circuit bent instrument, circa 1967–68. The original was destroyed by "an irate audience" during a performance. *Source*: Reed Ghazala, *Circuit-Bending: Build Your Own Alien Instruments* (Hoboken, NJ: Wiley, 2005), 8. Courtesy Reed Ghazala.

randomized results, immediacy, and the discovery of the undocumented or the concealed. The aesthetics of circuit bending are quirky, handmade, and one-of-a-kind, and practitioners value the entire process of developing and exploring, not just the end result.

The process of circuit bending typically involves going to a secondhand store or garage sale to obtain an inexpensive battery-powered device, taking the back cover of the device off, and probing the mechanism's circuit board. Any two points on the circuit board are then connected by using a "jumper" wire that temporarily short-circuits and rewires the device (see figure 2.1.2). Any conductive piece of metal will do; Ghazala used a bent hairpin in his early experiments. The battery-powered device is on during this process so the artist can hear the results of their probing, which is causing electric current to be rerouted in ways the device was not originally designed for.

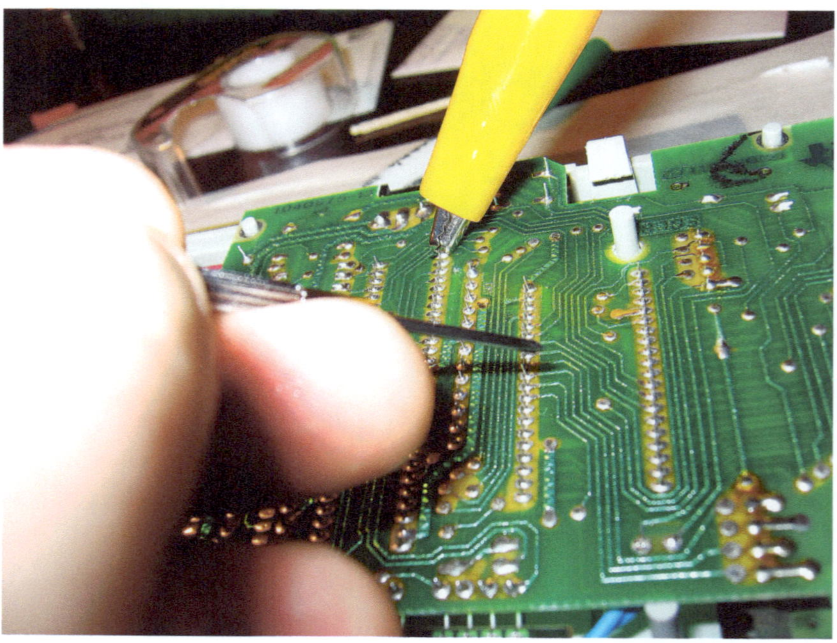

Figure 2.1.2
The process of probing for "bends," where a jeweler's precision screwdriver electrically connected to an alligator clip is used to probe soldered connections on the back of a circuit board. Courtesy Cole Moeller.

If an interesting result is found, the connections are marked for modification, and switches, buttons, or other devices can be inserted between these points to enable or disable the effect. Optionally, the connections can be modified to be sensitive to touch: conductive metal pieces can be soldered to the points as "body contacts," where the performer simultaneously touches both connections and has the low-voltage electricity flow through their body to enable the effect. This process of exploratory reverse engineering continues on a trial-and-error basis over the circuit board: probing, locating bends, and intervening in the circuit. The joy of discovering sounds and effects that are concealed in the system is part of the allure of the process, and when multiple bends are combined, the bent systems often become unpredictable, shifting and recombining. The end result is the handmade customization of a device that might have been considered obsolete.

It should be noted that several historical precedents for both musical chance encounters and electronic musical experimentation existed long before Ghazala, including those from people like Leon Theremin and Maurice Martenot—with their electronic gestural instruments—and John Cage, who began using chance procedures in 1951.[4] Nam June Paik famously bent the electrons of a tube TV in 1965 with Magnet TV.[5] Ghazala's unique contribution can be seen as combining these variables and articulating a methodology for them through disassembled postconsumer devices.[6]

Hacking Speak & Spell: The *Incantor*

Likely the most recognizable and iconic example of circuit bending is Reed Ghazala's *Incantor* series of devices, commonly known informally as "hacked Speak & Spells." Ghazala has been building these highly customized Speak & Spell, Speak & Read, and Speak & Math children's toys since 1978 (figure 2.1.3). As an outgrowth of Texas Instrument's research in the area of synthetic speech, the Speak & Spell learning toy was designed to educate children aged seven and older in how to spell and pronounce more than 200 commonly misspelled words.[7] However, Ghazala's *Incantor* completely reconfigures the synthesized human voice circuitry to spew out a noisy, glitchy, tangle of sound that stutters, loops, screams, and beats (figure 2.1.4). "One of my modified Speak & Reads, in a voice that sounds

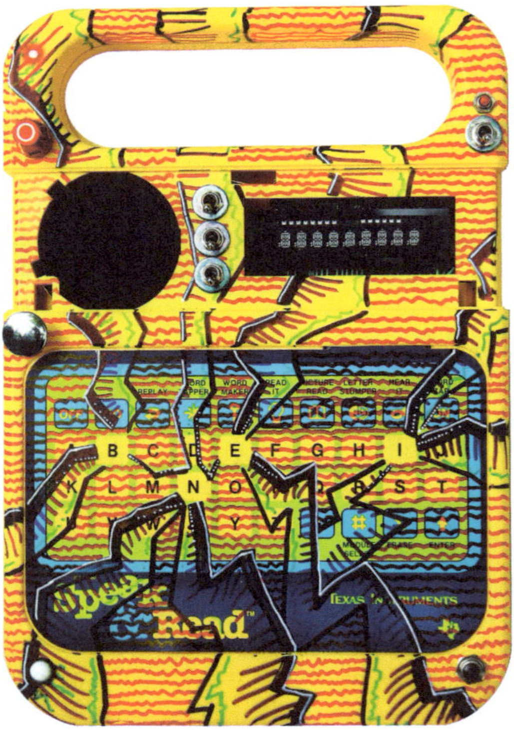

Figure 2.1.3
An *Incantor*, a modified, or "circuit bent" Speak & Read, by Reed Ghazala. First developed in 1978. Courtesy Reed Ghazala.

like an intoxicated Jack Benny, occasionally says 'Let's smell the scissors s'more' . . . [and some] of the sounds are indescribable, but an example might be: 'Ayy'—bell sound—pitch sweeps—'oh'—metal crash—bubble sound."[8]

Similar to the early twentieth-century artists who employed readymade materials, Ghazala sees the surplus of consumer devices as an "immediate canvas" that can be simply modified to discover a mysterious, surreal world of sound.[9] Although the junk culture of reuse is consistent between early twentieth-century readymade artists and contemporary circuit benders, there is a distinct difference in present-day work: bending takes a unique pride in the folk process of reverse engineering without formal expertise and prizes one-of-a-kind inventions. As opposed to Marcel Duchamp taking a machine-manufactured product and positioning it, almost sarcastically, as a handmade piece of art, this work rejects the machine-manufactured aesthetic of the source product and alters it into an idiosyncratic and crafted piece. The mechanically reproduced original device is taken from a generic form and hand tuned into a feral kind of invention, introducing a sense of aura, or a unique "presence in time and place."[10] This is achieved through the process of bending, a type of contemporary ceremony where a hidden underlayer is explored, perhaps reminding us of Walter Benjamin's belief that "the unique value of the 'authentic' work of art has its basis in ritual, the location of its original use value."[11]

Communities of Error

Circuit bending, as promoted by Ghazala and others, has developed over the last few decades into a diverse network of practitioners, with semiannual international festivals dedicated to the topic running since at least 2004.[12] The community of circuit bending feeds off of inexpensiveness and immediacy. In comparison to other electronic techniques, circuit bending is a simple and straightforward electronic audio design process that requires minimal prior knowledge of electronics. The continued proliferation of consumer electronics has made the raw materials of circuit bending ubiquitous, with folk knowledge around circuit bending practices increasingly dispersed over the Internet, in magazines, and through local hackerspace workshops (figure 2.1.4).

Critics of circuit bending, like Jeff Morton, raise several valid concerns about the scene. For one, only considering Reed Ghazala as the sole "father" of the field cuts off clear historical precedents, like Leon Theremin, Ondes Martenot, and John Cage. Although the term "circuit bending" is useful, it also contains some danger. Blindly adopting Ghazala's origin story can establish a cult-like scene with Ghazala at its core and limiting the field of practice to his style of work. One example of this influence can be seen in the similarity and even conformity of circuit bending events, workshops, and performances. As Morton outlines, "My work with circuit-bending has become refined to the point where I recognize similarities and idiosyncrasies in basically every electronic toy I encounter. There are many permutations and a practically infinite variety of interconnections, but it is always contained within the parameters of the original system."[13] Although the events celebrate aleatoric chance encounters, the formula is, ironically, predictable. Events often feature a table with a bunch predominantly white men poking around the circuits of broken-apart, battery-powered toys (figure 2.1.5).

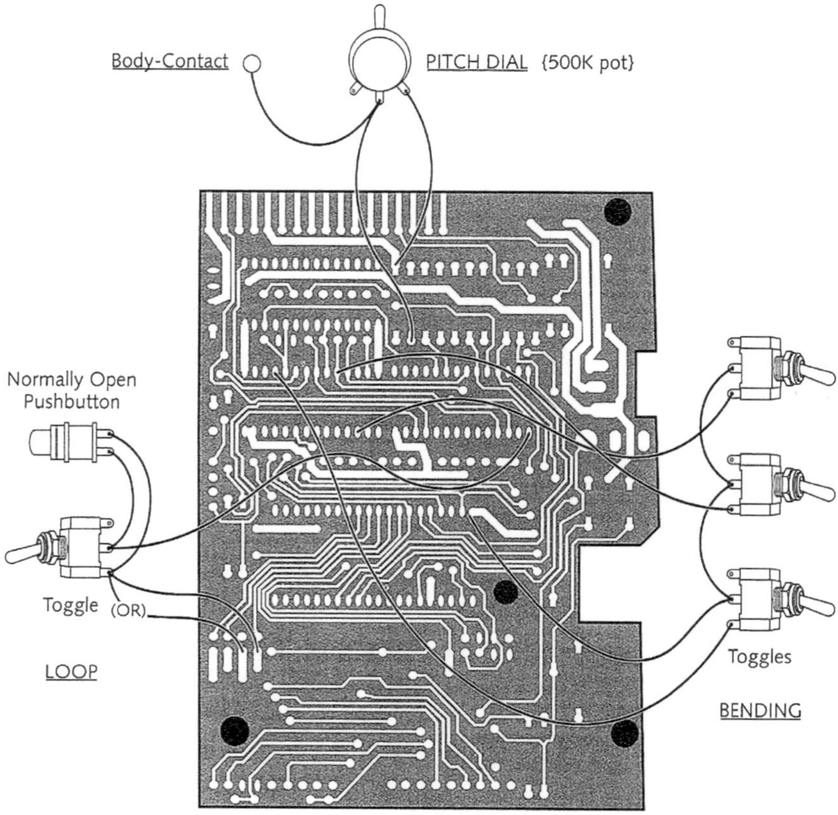

Figure 2.1.4
Circuit bending diagram for a membrane-keypad Speak & Spell circuit board that forms part of the Incantor. *Source*: Reed Ghazala, *Circuit-Bending: Build Your Own Alien Instruments* (Hoboken, NJ: Wiley, 2005), 219, fig. 15–2. Courtesy Reed Ghazala.

Demanding Depunctualization: Hackability, the Right to Repair, and Open Source Hardware

Circuit bending and similar practices take the trash of electronic culture as a starting point, and its key tactic of reuse is driven by a punk-like breaking apart and exploration of blackboxed consumer technologies. Sociologist and anthropologist Anne Galloway identifies that DIY practitioners have the ability to create their own worlds through DIY methods like circuit bending: "Similar ethics and practices can also be found in punk rock

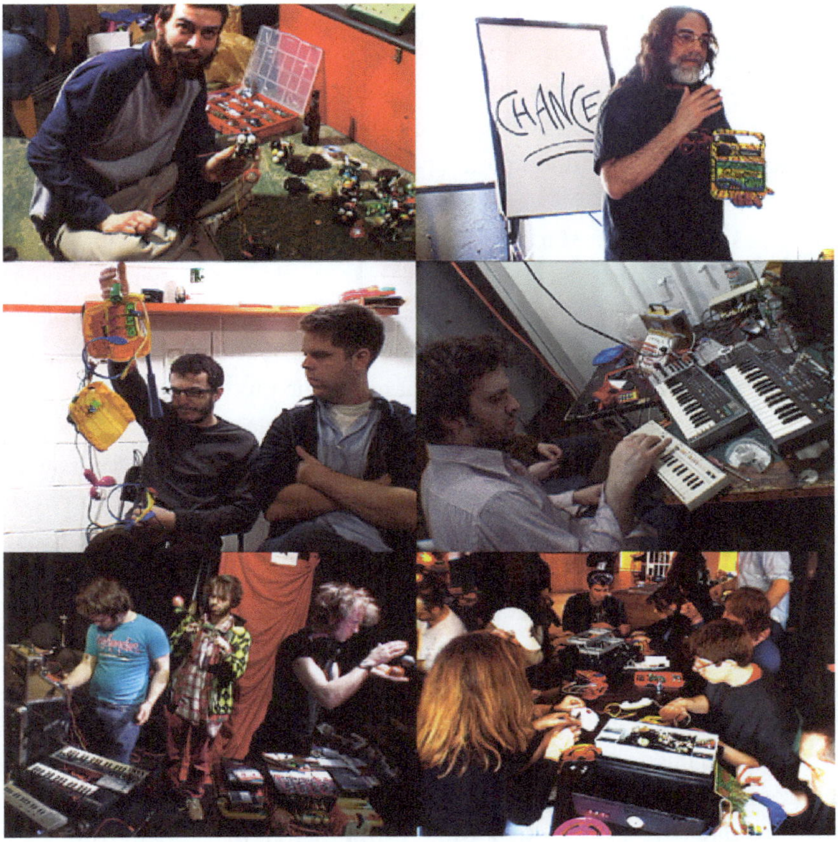

Figure 2.1.5
Back DVD cover image from *Bent,* Derek Sajbel's 2004 documentary about the first International Circuit Bending Festival (NTSC, 150 minutes). Courtesy Derek Sajbel.

'do-it-yourself' (DIY) cultures. The general premise behind DIY is that if you do not like the way things are done, then you should do it yourself. DIY culture involves creating your own world amid the dominant culture, thereby putting power back in the hands of individuals."[14] DIY culture uses available, inexpensive materials to shape one's own cultural identity by building what one feels is missing from the mainstream—your own album, your own book or magazine, your own clothes, or your own musical instrument.[15] "Around the world, people are making things themselves in order to save money, to customize goods to suit their exact needs and interests, and to feel less dependent on the corporations that manufacture and distribute most of the products and media we consume."[16]

However, for circuit bending and DIY practices involving information technology, a problem lies with the blackboxing and planned obsolescence of technology. Contemporary methods of manufacturing and planned obsolescence result in devices that are not only difficult to modify, but also intentionally engineered to be difficult to modify; although the devices are inexpensive, they are inaccessible, and their interiors are generally the territory of experts. The affordance of user serviceability in many electronic devices has been systematically removed to save space, weight, or cost, or to encourage users to replace instead of repair the products.

Demanding serviceable or modifiable technologies is not only an effort to lengthen the projected lifespan of consumer products, but is also viewed as a deeper issue of ownership and of blurring the demarcations between consumers and producers. This demand has been termed by some authors as an appeal for "hackability"—for consumer devices that are depunctualizable, explorable, reconfigurable, and customizable.[17] Hackability is also intertwined with the important topic of repairability, which can be seen as a legal, technical, and environmental issue through governmental initiatives like the "Right to Repair" bills. Politicians like Massachusetts Senator Elizabeth Warren have proposed legislation to protect farmers from being bound to proprietary repairs for agricultural equipment: "Newer John Deere equipment comes loaded with software and firmware that make it impossible for farmers to fix their own equipment. . . . Broadly speaking, right-to-repair legislation would have a measurable impact on the average consumer's life, from the farmers hacking their tractors to people suffering from sleep apnea who tweak their CPAP machines to folks who just want a decent price on a replacement iPhone screen."[18] In summary, the blackboxing of technology can be defined as "a process that makes the joint production of actors and artifacts entirely opaque," and hackability can be seen as a desire to make product actors and artifacts more visible and revealed.[19]

A key component of electronic DIY culture relates to the values of open-source software and to extending these values to manufactured electronic hardware devices. In the open-source mindset, a hackable device has fewer corporate dependencies when it fails. Systems are better when they are technically accessible and nonproprietary, and when they include relevant documentation to assist in the modification of the device. Unlike computer software, no End User License Agreement (EULA) for hardware usually exists; instead, the device is physically constructed with technical barriers that limit nonexperts from accessing the mechanisms inside of it.[20] Ed van Hinte envisions this as two territories of ownership of a device: the exterior skin of a product that is touched and owned by consumers, and an interior that belongs to somebody else. "This segregation into territories is quite peculiar. The person who buys a vacuum cleaner or a walkman does not have the entire purchase at one's disposal. Yes, you can unscrew it and open it up—if you understand how to do this, that is, which might be quite a puzzle—but your attempt will be punished by loss of your guarantee, and in the case of some larger appliances you may even be threatened with death."[21]

Taking the lead from open-source software advocates, *Make* magazine and other organizations position hackability as an issue of consumer rights and ownership.[22] Specific design and manufacturing processes—like glued-shut cases, the ease of repair of devices, a lack of documentation, the requirement of proprietary tools for repair, and proprietary batteries or chargers—are violations of consumers having outright ownership of their hardware (figure 2.1.6). This concept of ownership can be summarized by the slogan "If you can't open it, you don't own it": when products break down, hidden power structures, dependencies, and infrastructures are revealed. To use Heidegger's terms, dependencies come forward when an object shifts from being ready-to-hand to being present-at-hand.[23]

Breaking apart and getting inside a piece of hardware usually shows a network of dependencies that tie the owner to the corporation that manufactured the device. Thacker and Galloway state that "you have not sufficiently understood power relationships . . . unless you have understood 'how it works' and 'who it works for'. . . . It is not only worthwhile but also necessary to have a technical as well as theoretical understanding of any given technology."[24] A hackable device, in this sense, has fewer dependencies than the average black box when ripped apart, or "depunctualized."

Bent Conclusions

As a theme, a significant cluster of DIY electronic practices focus on unpacking and exploring the "black boxes" of technology and changing the taken-for-granted function of the technology without formal training

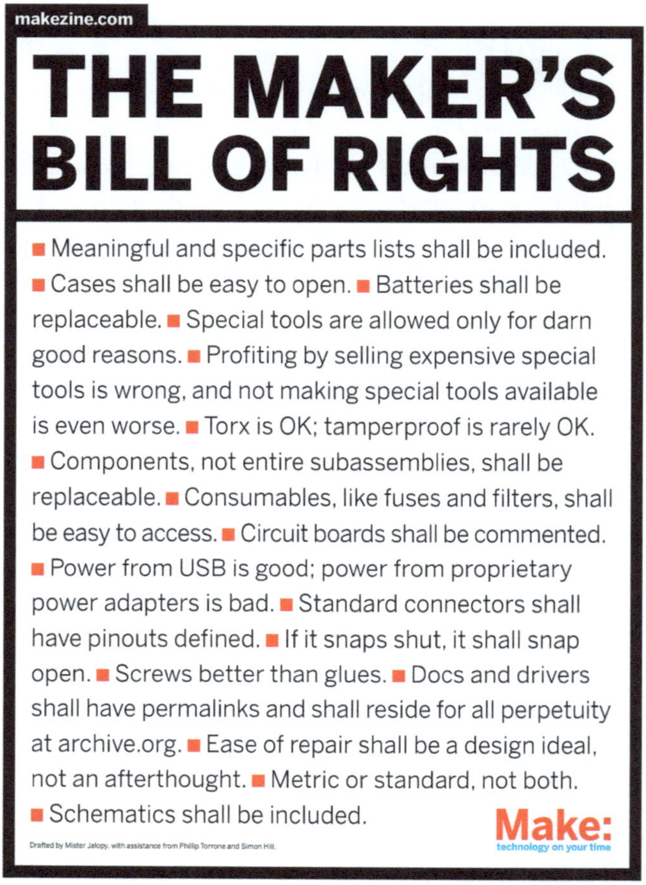

makezine.com

THE MAKER'S BILL OF RIGHTS

■ Meaningful and specific parts lists shall be included. ■ Cases shall be easy to open. ■ Batteries shall be replaceable. ■ Special tools are allowed only for darn good reasons. ■ Profiting by selling expensive special tools is wrong, and not making special tools available is even worse. ■ Torx is OK; tamperproof is rarely OK. ■ Components, not entire subassemblies, shall be replaceable. ■ Consumables, like fuses and filters, shall be easy to access. ■ Circuit boards shall be commented. ■ Power from USB is good; power from proprietary power adapters is bad. ■ Standard connectors shall have pinouts defined. ■ If it snaps shut, it shall snap open. ■ Screws better than glues. ■ Docs and drivers shall have permalinks and shall reside for all perpetuity at archive.org. ■ Ease of repair shall be a design ideal, not an afterthought. ■ Metric or standard, not both. ■ Schematics shall be included.

Drafted by Mister Jalopy, with assistance from Phillip Torrone and Simon Hill.

Make: technology on your time

Figure 2.1.6
Make magazine's Owner's Manifesto, a bill of rights to accessible, extensive, and repairable hardware. *Source*: *Make* magazine issue 4 (2005). Courtesy Mister Jalopy / Peter Vermeren.

or approval. Circuit bending, then, is an electronic DIY movement focused on manipulating circuits and changing the taken-for-granted function of the technology without formal training or approval. In de Certeau's terms, "These 'ways of operating' constitute the innumerable practices by means of which users reappropriate the space organized by techniques of sociocultural production."[25]

In the practice of opening up black boxes, error and malfunction typically spill out, but these malfunctions can be bent, transformed, and amplified into features. In other words, a black box is opened and broken apart, producing components that are disorganized, jumbled up, and without an explicit use, which then results in errors and trash that are creatively twisted into use.[26] Circuit benders work with the exploded fragments of broken technologies, in other words. These fragments can be more pleasing to the creator than the original role of the device, and in the case of circuit bending, errors become the prime feature of the modified device. Error—in some ways—is the willingness to let materials and technologies play their own role or to have efficacy in the creative process.[27] As Greg Siegel discusses in *Forensic Media* on the topic of black boxes, cybernetics and communication theory have historically viewed noise as something to be canceled—although the invention of the flight voice recorder changed that. Often *noise* is the most information-loaded component recorded when catastrophe hits.[28]

In summary, circuit bending is a way of operating that reminds us that users consistently reappropriate, customize, and manipulate consumer products in unexpected ways, even when the inner workings of devices are intentionally engineered to be the zone of experts. Reed Ghazala's *Incantor* is useful as a tool to remind us of sociotechnical issues in contemporary society, including planned obsolescence, the blackboxing of technology, and the interior inaccessibility and ownership of everyday consumer products. As a way of operating, circuit bending is an aspect of digital culture that does not fit under the term "new media" or in standard frameworks of innovation. The customized and folksy DIY methodologies of circuit bending recall a pre-new-media attitude of handmade craft and a post-new-media recycling of surplus electronics. DIY electronic practices like circuit bending clearly bend more than electronic circuits; they creatively prompt us to rethink the sociotechnical issues of planned obsolescence and the complex layers of ownership within digital culture and artifacts.

2.2 Exploration and *Wire Figures*: Technologies, Interactivity, and Radio Shack Cybernetics

DIY electronic artists often do more than breaking open and bending existing circuits into customized devices—they regularly build their own systems from scratch. Because most consumer technologies are black boxes with concealed functionality, building from scratch might be thought of as putting together a transparent or "white" box that highlights the interior functioning of a system. This goal of building for technological transparency shares kinship with movements like the open-source community and with circuit benders like Reed Ghazala. All are driven by the goal of unblackboxing complex technological systems to make a show of the working components. In this chapter, the black boxes of hidden technologies are not only opened and bent—they are completely rebuilt into systems that strive to be metaphorically and physically transparent. This "transparent boxing" approach emerges as an essential tactic used by DIY electronic artists, designers, and hardware hackers.

Transparent Boxes: Norbert Wiener, *Cybernetics*, and Interactivity

Norbert Wiener, in his classic book *Cybernetics* from 1948, defines a black box as a piece of equipment "which performs a definite operation on the present and past of the input potential, but for which we do not necessarily have any information of the structure by which this operation is performed."[1] In other words, it is an opaque object that processes information in a mysterious interior. For most people, a mobile phone serves as a clear contemporary example of a black box. Wiener contrasts this to a white box with an interior that has "a definite structural plan," like a well-documented open-source project with accessible code. Within the context of communication theory, these box metaphors were used to describe the science of transmitting messages between humans and machines, or from machine to machine. Wiener's point in *Cybernetics* is that animals, humans, and machines are not closed systems. Rather, they have complicated networks of feedback, communication, and control that span individuals and machines. From the vantage point of the first generation of cybernetic theory, animals, machines, and humans are equal in terms of being able to communicate and modify their own behavior through the feedback of information.

Wiener envisioned a world where machines were more physiological and "alive" through feedback, control, and self-regulation. He proposed steam engine governors, thermostats, automatic gyrocompass ship-steering systems, target-seeking self-propelled missiles, and computing machines as examples of technologies that displayed an awareness of their environment through self-regulation.[2] A shared understanding of information feedback in early cybernetics helped form concepts of interactive technologies and the idea that even relatively simple electrical systems could display complex emergent behaviors. Consequently, this first generation of cybernetic theory was essential for society to understand information as transmittable and malleable. In this way, the field of cybernetics helped shift society's perceptions from a more basic "on-off" world of *electrical* buttons and switches to the information-processing world of *electronic* systems.

Influenced by cybernetics, designers and artists in this period shifted into the mindset of building interactive electronic systems and designing behavioral processes. Examples from the late 1950s include the cybernetic sculptures of Nicolas Schöffer and music compositions by John Cage and Iannis Xenakis.[3] In the case of Xenakis, part of his working process of writing music consisted of taking cybernetic methodology—W. Ross Ashby's *An Introduction to Cybernetics* from 1956 for example—and applying it step by step to produce what he termed "Markovian stochastic music" work, or notes generated through the laws of probability where subsequent notes are partially determined by their predecessors.[4] Cybernetics opened a door to understanding that information was a thing that could sculpt, shape, and modify things on its own. Interactive systems were not just reacting to user input, but were capable of modifying themselves as a nascent form of artificial intelligence.[5]

This influence of cybernetics spread through academic and popular culture in the 1960s. Jack Burnham's essay "Systems Esthetics" in the September

1968 issue of *Artforum* marked a highpoint of this mindset within the field of art.[6] However, for amateur technologists, the complex systems described by Wiener and others were difficult to build. Consequently, electronic systems—and systemic approaches—were primarily the territory of professional engineers, scientists, and the military.

Radio Shack Cybernetics

American hobbyists interested in experimenting with cybernetic concepts often started at places like Radio Shack, a then-popular (and now defunct) American retail store that catered to the hobby electronics market. In addition to carrying amateur radio kits, consumer electronic products, and electronic parts, Radio Shack carried a number of printed DIY guides that served to educate amateurs. In the 1970s they started printing their own guides, with author Forrest Mims publishing an impressive total of thirty-six books for the franchise between 1972 and 2003. These books included hand-drawn circuit diagrams and hand-lettered text and were written as approachable electronics guides for amateurs (figure 2.2.3).[7]

Although many artists in the 1970s and 1980s knew of cybernetic theory, their actual experimentation with electronics typically began in stores catering to DIY electronics. A good example of this is the Canadian artist Diana Burgoyne, who started working in "cybernetic art" in 1982. She primarily used circuits described by Forrest Mims and parts from Radio Shack and other similar suppliers to make complex, interactive works of art.[8] In this way, Burgoyne embodies a type of "Radio Shack Cybernetics" approach that uses mass-market parts and DIY tutorials to build interactive artworks.

Wire Figures is a sculptural installation built by Diana Burgoyne in 1985 that consists of two human figures: one male figure and one female figure primarily constructed out of bent copper wire (figure 2.2.1). The copper wire is approximately the thickness of a coat hanger and curves in and around on itself like a meandering three-dimensional doodle. Included among the pair of serpentine human silhouettes are several speakers and raw electronic components, somewhat reminiscent of Paik's *Robot K-456*, although Burgoyne's work is considerably more refined and is described by Burgoyne as handcrafted electronic folk art.[9] The two figures are installed in a gallery space a few meters away from each other, and each body emits sound when it is physically interacted with by the audience.

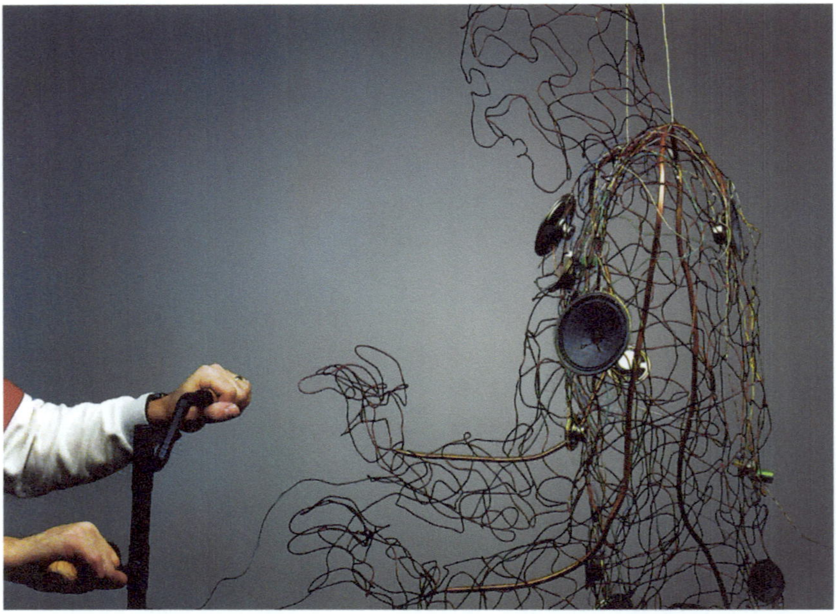

Figure 2.2.1
Diana Burgoyne's *Wire Figures* (1985). Audience members can grasp the handlebars (left), which triggers audio components in the electronic sculpture (right). Courtesy Diana Burgoyne.

The male figure has a set of handlebars in front of it that trigger sound when grasped (figure 2.2.2). With wire coiled around the grips, the grips attach to a vertical pole with plumbing solder. When holding the grips, the gallerygoer stands face-to-face with the male figure. When viewers grasp the handlebars, the sculpture emits a swooping electronic pitch that quickly climbs the harder the handlebars get gripped. From a technical perspective, the project works by measuring how much electrical resistance exists between the handlebars—in other words, how easy it is for electricity to flow from one hand to the other. When a person grasps the handles, the person's entire body changes the electrical resistance of the circuit. This level of resistance modifies an integrated circuit timer chip that generates a square wave amplified as audio out of multiple speakers positioned around the wire framed body. The tone starts at a growly low note—approximately 1,000 hertz—and as the grip tightens the tone rises into a 15,000 hertz screech. The harder the handlebars are clutched, the contact increases and

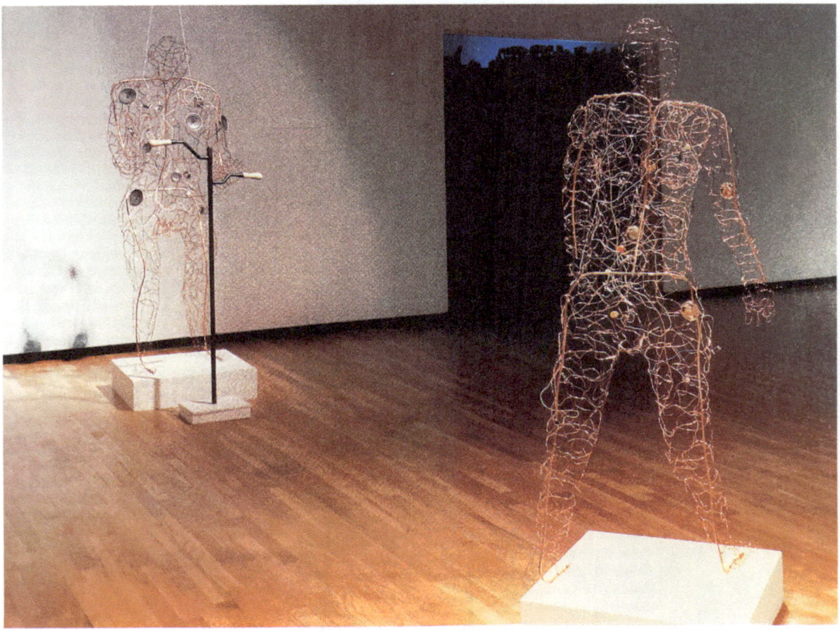

Figure 2.2.2
Diana Burgoyne's *Wire Figures* as installed at the Walter Phillips Gallery in 1987 for the exhibition titled "Siting Technology" in Banff, Alberta, Canada. Courtesy Diana Burgoyne.

the resistance goes down, causing the tone to go up. The circuit design came directly from Mims's guidebook to electronics sold at Radio Shack, and it used approximately $20 in components.[10]

The female wire figure has light sensors that activate sound when individuals cast shadows on the piece—a "dark trigger relay." Built with similar materials, this figure is also built with irregularly coiled copper wire forming a full-scale human. The sculpture has no handlebars but has four light sensors positioned around the body, with the piece intentionally lit so that viewers' shadows intermittently darken regions of it as people look at the piece. Each of the four light sensors connect to a sound chip that triggers amplified sound when the light decreases, emitting a playful assortment of electronic sounds. Like the male circuit, this circuit design also came from Forrest Mims's guidebooks from Radio Shack and was inexpensive to produce (figure 2.2.3).

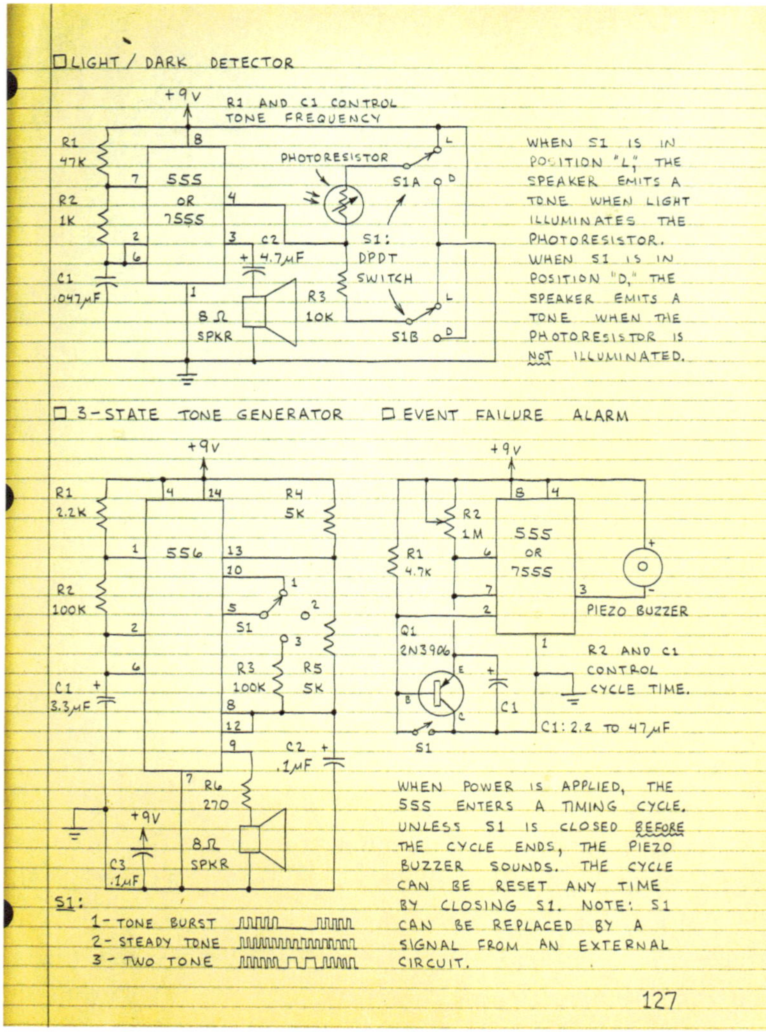

Figure 2.2.3

Like several other artists and hobbyists, Diana Burgoyne based *Wire Figures* on circuits from Forrest Mims's books on electronics that were widely distributed by Radio Shack. In particular, Burgoyne's female figure used a modified version of the "Light / Dark Detector" circuit (top). Mims's guides were unique in that they were often formatted as an engineer's handwritten notebook with informal notes and drawings. Courtesy Forrest Mims.

Wire Figures shows how a classic do-it-yourself guide with simple electronics can be scaffolded into complex sculptural projects. Burgoyne's work reminds us that technically straightforward and relatively inexpensive technologies can be made into complex artworks through sculpture. In other words, Mims's Radio Shack circuits were the starting point for the lights, the space, and the audience interaction with the project that formed the actual "piece."

Demystifying Technology through Folk Construction

Burgoyne started her artistic career by studying sculpture at the University of Victoria with recognized sculptors Roland Brener and Mowry Baden. Her initial introduction to electronics was through two graduate students—Doug Collins and Steve Parker—doing theater work in the late 1970s using large, floor-mounted pads that had electronic components in them. Burgoyne later studied at the University of California, Los Angeles (UCLA) with Chris Burden and Charles Ray. As Daniel Jolliffe has outlined, these artists had direct overlap with Burgoyne's time-based sculptural practice—"in Mowry Baden's kinaesthetic awareness of the body, Brener's discipline in the formal construction of objects, Ray's canny sense of objects as they relate to the body, or Chris Burden's startling execution of the performative act."[11] Burgoyne takes these influences from time-based performative sculpture and merges them with handmade component-level electronics. These handmade circuits primarily output sound and are combined with performance and installations with the intention of raising questions about the role of technology in society. In Burgoyne's wider body of work, technology is seen as dehumanizing by default and is rescued through exploding, reassembling, and reconfiguring it into a more empathetic program by building handmade systems that the public can view and interact with.

Wire Figures and other DIY electronic art projects work to demystify technology. *Wire Figures* strives to make the "black box" of technology into a transparent (or semitransparent) box in several ways. First, it makes the electronic technology visible in its physical form, laying out the raw circuit in front of viewers instead of sealing it up or hiding it away. To use a box metaphor, the project has no exterior shell. Second, the project humanizes the technology through the construction of life-sized human figures, which creates more empathy in viewers than if the work was in the form of an abstract shape. Third, the project demystifies technology by creating a system that invites playful interaction. The mechanism of triggering the sounds in the female figure, for example, is not visible to most viewers, which results in the audience trying to trigger the audio by moving around the space, flapping their arms, or hovering their hands over the body's surface. The grip triggering the audio in the male figure also works with two or more people, as long as both of the handlebars are touched and the people hold hands to form a loop. Because of this, the artwork can create situations where a large circle of people can form a ring with the handlebars as a link, and the audio can be modified by how hard they squeeze each other's hands, kiss, or engage in other forms of physical contact. Artwork like *Wire Figures* can be an inspiration for socializing and experimentation, bringing people into new circuits of communication.

This type of literal exposure of electronics can be seen in many other artworks as well. Examples include Canadian artist Ken Gregory's *Electronic Mail Cards* (2003), where postcard-sized circuits that synthesize and play sounds are sent through the postal system. These circuits are constructed with bare electronic components stuck to a piece of cardboard, featuring a computer (the BASIC Stamp microprocessor), support electronics, an audio speaker, batteries, and a miniature LCD display that can display messages. The project is a useful example of how both technology and its distribution can be made more transparent with DIY methods: Gregory's dissemination is publicly distributed in a style reminiscent of Fluxus, with no special equipment or expertise required, and no need to go into a gallery space to experience it.[12]

Another key component of Burgoyne's artwork and other DIY electronic projects is that they highlight how unimaginative interactivity and control often are when embedded in mainstream electronics. Burgoyne summarizes this motivation as follows:

> Making technology and the viewer interact so that the viewer becomes the performer and, as a result, controls the environment is an important aspect of my work. With this series, my objective has been to humanize a material which many people find intimidating and dehumanizing. The electronic circuits create sound, light or movement, and by placing the viewer in a position of control, it takes them out of their passive role and places them in a very active one . . . setting up relationships which raise broader questions about human co-existence with technology.[13]

Wire Figures opens up technology, makes it visible, and enables users to enter into an open-ended play scenario with the flow of electricity that is generally hidden in a black box. Through interactivity and visibility, Burgoyne's project works to demystify technology.

Transparent Conclusions: The Humanization of Control through Participatory Systems

The humanization of control, then, is pivotal in Burgoyne's *Wire Figures*. This humanization of control—and using electronics to accomplish this—is a core theme that cuts across a wide swath of electronic DIY projects in the domain of art. This process of humanizing control is often the nucleus of interactive art, where artists are more concerned with sculpting interactive systems and behaviors than objects. Burgoyne's artwork humanizes technology with two core techniques: inviting tangible interaction from gallery viewers, and creating lovingly handcrafted constructions. By putting the audience in control of a visible, handcrafted interactive system, Burgoyne illustrates how cybernetics can be performed, increasing transparency in electronic art.

Jack Burnham, in his 1968 *Artforum* essay titled "Systems Esthetics," foresaw the impact of cybernetics and flagged this as a major shift in contemporary art, claiming that "we are now in transition from an object-oriented to a systems-oriented culture. Here change emanates not from things but from the way things are done." Systems- and interactive-oriented work emerged out of the field of cybernetics, which was pivotal in understanding concepts like feedback, control, and shifting the field of on-off electrical systems into information-processing electronic technologies.

This style of direct interactivity—where audience members trigger a circuit that gives a direct and proportional response in return—can be thought of as a "first order" type of interactivity. In the case of *Wire Figures*, touch and light trigger sound in a proportional way. In this first order of interactivity, spectators are required to be participants in order for projects to be fully functional. In the case of the male figure in *Wire Figures*, touching the handlebars connected to the figure's circuit emits a sound, and the harder the participant grasps, the higher the pitch. This emphasis on participation instead of viewing in a gallery setting shifts the experience from art as a visible object into art as a responsive environment and experienced "happening." The aforementioned Fluxus happenings and work done through groups like Experiments in Art and Technology are without reservation relevant here.[14]

By the time of Burgoyne's *Wire Figures* in 1985, interactive electronic systems were a relatively well-understood path within the field of contemporary art, with authors like Frank Popper writing about the topic since at least 1968.[15] However, although this path was understood, it was not regularly traveled. Popper, like Burnham, tracks that this trend is part of the larger movement away from things and into systems: "The increased participation of the spectator has been one of the factors that has brought about the disappearance of the traditional art object. . . . The emphasis lies not on the object, but on the dramatic confrontation or the perceptual situation."[16]

It is worthwhile to note that physical transparency and interactivity do not provide a didactic explanation of how the technology works. Although a component is put into the open, its exact function remains a mystery. In this way, Burgoyne's transparency is more of a window into an electronic technology than a practical tutorial about how it functions. It invites participation through physical touch and exposure of functionality, not by explanation.

Some theorists like Varvara Guljajeva have classified Burgoyne's direct style of interactive work as being "participatory." This is in contrast to Guljajeva's category of "post-participatory," which is more meta-interactive by manipulating data instead of having direct user input. In other words, projects like *Wire Figures* would be classified as participatory because they have a direct "this-does-that" system of control: pushing a button gives a response, more or less. Post-participatory systems have "no audience involvement, but still incorporate an internal system interaction with a data source"[17]—instead of human–machine interaction, there is more of a system-to-system type of interaction. A historical example of this is Nancy Paterson's *Stock Market Skirt* (1998), an interactive project that robotically raises a skirt's hemline based on stock market prices, where the higher the stock, the shorter the dress. Both *Wire Figures* and *Stock Market Skirt* are interactive, but the former directly reacts to people and the latter reacts to (or renders) an information flow. Instead of post-participatory as a label, alternate terms might be "meta-interactive" or "data-interactive."

In summary, Burgoyne's work provides a compelling example of how relatively simple and low-cost technologies can shift the gallery space from an observational and object-oriented place into a more interactive and system-oriented framework. *Wire Figures* also shows the impressive ingenuity of the DIY mindset. Inexpensive parts and plans from Radio Shack are blended with relatively complex concepts from cybernetics to build responsive and interactive sculptural work. The end result is considerably more poetic, mysterious, and playfully engaging than what is commonly found in consumer culture. In the process, Burgoyne opens up the black box of technology and presents an elegantly transparent system that teaches us part of the complex dynamics and logic of the DIY mindset.[18]

2.3 Exploration and *20 Oscillators in 20 Minutes*: Technological Performance, Hedonization, and the Thrill of Impending Failure

Inspired by the DIY radio-craft culture of the early twentieth century, Darsha Hewitt's work takes on an engaging workshop-style tone that is explorational, educational, and conversational. As a multidisciplinary artist that works with technology, Hewitt primarily works with handmade audio electronics through electromechanical sound installations, experimental performance, drawings, and videos. An example of this is *20 Oscillators in 20 Minutes*, a technically challenging, time-limited performance in which Hewitt scrambles to build and use twenty square wave oscillators in real time (figure 2.3.1). She essentially fabricates a twenty-track miniature drum machine from scratch using bare electronic parts—and she makes a hilarious and exciting slapstick technical performance of it in the process. It is simultaneously a technical demo in electronics, a conceptual musical performance, and an edge-of-your-seat hackersport.

The Handmade Electronics of Darsha Hewitt

The project is built on electronics breadboard, typically used to prototype circuits due to its ease of plugging in wires rather than soldering to make connections. Each oscillator consists of a 555 timer chip, two capacitors, two resistors, and four pieces of wire. Hewitt assembles one oscillator at a time, plugging each component into the breadboard piece by piece. The terrain of the circuit is built up during the performance, physically resembling a miniature alien landscape inhabited by an outcropping of black silos, eight-legged beings, striped logs, and numerous Hula-Hoops of various sizes lodged halfway in the ground.

The piece—which was first performed in 2012 at the New Music Festival in Vancouver, Canada—starts with Hewitt putting her electrical components

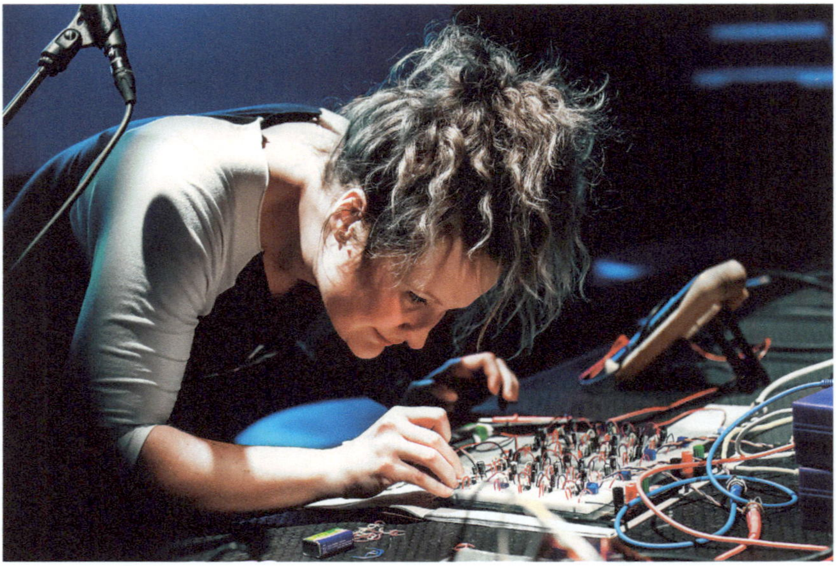

Figure 2.3.1
Darsha Hewitt performing *20 Oscillators in 20 Minutes* at the Piksel 2012 Festival in Bergen, Norway. In the project, an audio circuit is assembled live in front of an audience. Courtesy Sergey Dushkin.

and parts on a table and plugging the circuit into a set of speakers. A twenty-minute timer begins, and Hewitt starts methodically assembling the oscillator circuits by connecting wires to electronic components she has prepared on the breadboard. While Hewitt is working, she casually talks to the audience, partly to fill the awkward silence before the circuit starts to generate any significant sounds. "Has anybody built electronics before? Has anybody built electronics in front of this many people before? Yeah, so you'd better fucking be nice."[1] The audience laughs along, occasionally responding to her tech stand-up. Through her conversation, Hewitt works to demystify the electronics, explaining what timer chip she is using, how she is nervous, and how she is just an ordinary person that is putting together some electronics under pressure in front of an audience. Her conversational tone is half stand-up comedian and half electronics instructor: "Anyone in the audience from Norway?!" is followed by "Anyone know what chip I'm using here?" and a quick description of the 555 microcontroller timer chip. Her tone is unpretentious in almost a cliché of a Canadian, where she quickly breaks down the "fourth wall" between herself and her observers, emphasizing her status as a peer instead of a performer. Depending on her venue, there is no fourth wall: the audience is often standing directly beside and around her, watching her scramble her circuits together out of raw components (figure 2.3.2).

After a short time, the first oscillator starts, emitting a heartbeat-like electronic "thump-thump," and as the system is assembled, more heartbeats are layered into the soundscape. As time goes on, she intermittently talks to the audience and tells stories of technological malfunction: "And then one time the solder fell in my lap and burned through my crappy jeggings. Who knows what jeggings are? Whooo!!"

As time continues, it becomes clearer that the goal of building twenty oscillator circuits in twenty minutes is substantially difficult. Errors and inconsistencies in how Hewitt has joined the circuits are audible in some of the oscillators emitting soft warbling or clicks—the circuit that Hewitt is working with gracefully degrades and spills an assortment of sounds even when not plugged in exactly as her reference oscillator. "Eleven minutes left? Shit, I dunno if I'm going to make it this time!" Hewitt is flustered, talking,

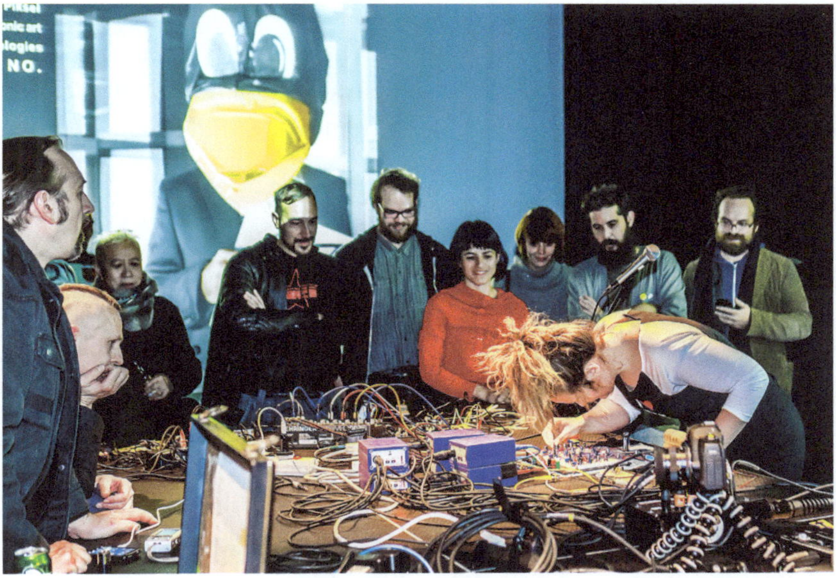

Figure 2.3.2
Darsha Hewitt performing *20 Oscillators in 20 Minutes* while interacting with the audience at the Piksel 2012 Festival in Bergen, Norway. Courtesy Sergey Dushkin.

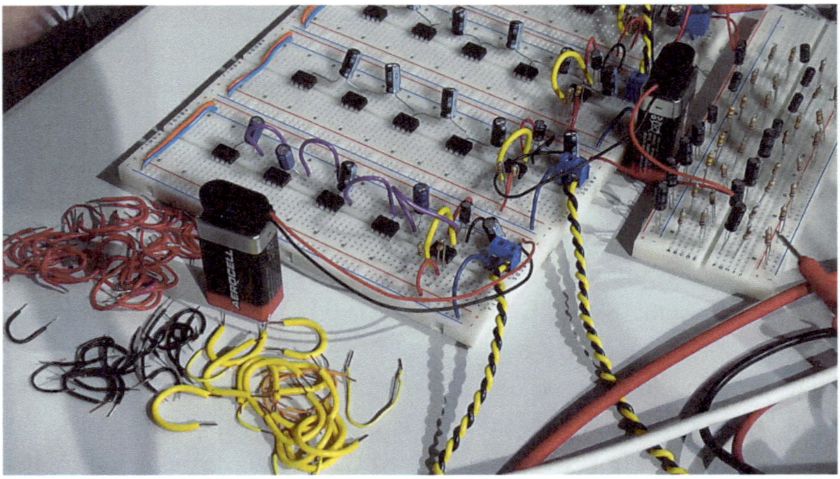

Figure 2.3.3
A detail of Darsha Hewitt's *20 Oscillators in 20 Minutes*, which is built on a solderless prototyping breadboard with electronic parts. Courtesy Sergey Dushkin.

sweating, and jamming away at her work, and the audience is slowly drawn in to cheer her on as the clock counts down to zero. As the clock strikes, she lifts her hands up in the air like a rodeo cowboy that has just roped a steer, with the audience cheering her accomplishment—although they're not exactly sure of the consequence of the impromptu electronic sports comedy.

Live Hardware and Exploratory Technological Performance

The live and performative assembly of hardware circuits is a process that has been explored by others, including Tetsuo Kogawa's performances building FM radio transmission equipment in the format of a television cooking show.[2] Similarly, several artists working in the sphere of electronic audio have performative circuit creation as a part of their work. For example, artists like John Bowers do work involving submerging a random pile of electronic components in a bowl filled with mildly conductive fluids like beer, saline, or blood. A low voltage is put through the clutter, and as different components settle on each other or are mixed, the system outputs a relatively dynamic electronic noise to speakers—a bit of a ready-made noise circuit mixture.[3] The Brooklyn-based collective Loud Objects, composed of Tristan Perich, Kunal Gupta, and Katie Shima, also perform live circuit construction music where they solder together audio circuits on top of a lit overhead projector (figure 2.3.4). The audience looks on as the group solders together a number of small, black microprocessors, wires, and other components: "The audience must wait for the process of connections which activates the circuit and awakens the sound within it to reach completion before their eyes."[4]

A similar fabrication-oriented project is Katherine Moriwaki and Jonah Brucker-Cohen's *Scrapyard Challenge Workshops*, which are intensive workshops where participants build electronic projects using found or discarded "junk."[5] Participants build simple electronic projects using hacked-apart old electronics, clothing, furniture, outdated computer equipment, appliances,

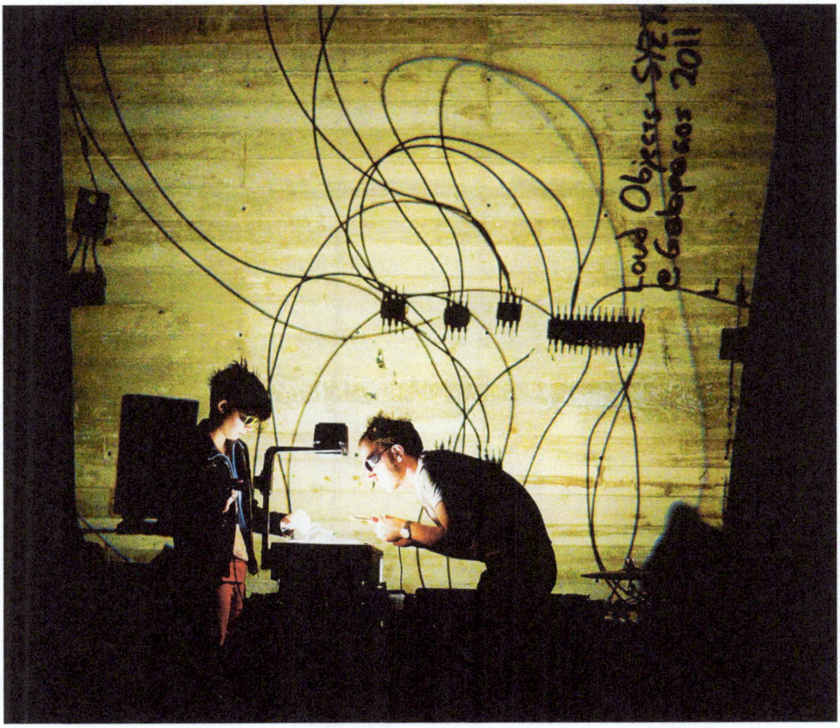

Figure 2.3.4

Loud Objects is a group with members Kunal Gupta, Tristan Perich, and Katie Shima that create custom audio circuits with simple electronic elements as a live performance. Courtesy Katie Shima.

turntables, monitors, or gadgets. Moriwaki and Brucker-Cohen's approach embraces ad hocism and jugaad approaches to deliver a memorable educational learning experience. Leveraging creative self-expression, the workshops generally conclude with a collaborative performance where people show their DIY technical creations. The project is highly successful in introducing students to electronics and interaction design, and it has been held hundreds of times in fifteen countries on five continents across the planet. *Scrapyard Challenge Workshops* clearly remind us that the exploratory and performative DIY space between art, design, and technology is incredibly useful as a tool for informal science education.

Live Coding as Exploratory Technological Performance

The workshops and processes above share some similarity to the practice of live coding, which is live computer programming as a type of artistic performance. Live coding regularly involves an individual who types out programming source code line by line on a laptop that is connected to a data projector—and the outputs of the compiled source code typically include image-based digital media or computer music.[6] The format is both educational and improvisational: both code and output are visible at the same time. Educational equivalents exist in most computer science programming classes, with professors regularly writing out code, compiling it, and showing what happens when portions are changed. From an improvisational perspective, the practice is common in computer music circles where software tools and algorithmically generated audio is performed. The practice exists as a workshop-performance that teaches, entertains, and reinserts a live human element into relatively technical fields.

Another performative and workshop example includes the work of Joel Ong. In June 2015, Ong presented a workshop titled *Those Who Observe the Wind,* which incorporates sound, sculpture, and data-triggered electronics.[7] The project was built over the course of two consecutive weekends where sighted and visually impaired students in the Seattle region worked together in workshops to build their own light-activated musical instruments. The instrument Ong designed for this workshop has a light-sensitive plucking mechanism and a hand-operated wheel to control the tuning of the pitch. Through this project, students learned the basics of physical computing and interacting through the Arduino and a light-sensitive photoresistor, and they also got hands-on experience with tuning and playing a stringed instrument.

In comparison to live coding, workshop, or live hardware assembly projects, however, Hewitt's performance is considerably more of a race against the clock and a test of speed, focus, and skill. Hewitt's *20 Oscillators* is a bit less about exploring the creative potentials of the circuit or code. The strict time limit flips it into a type of sporting event that pulls the audience in, especially if failure is imminent: "People seem to lose their mind in the right kind of way. It's pretty fun to see art people at a serious experimental music event get worked up like they are at a sporting event."[8] Untimed live coding or hardware construction performances, in contrast, are more exploratory—perhaps more similar to a musical performance like Bamanya Brian working his "Afrorack" DIY modular analog synthesizer by tweaking knobs, patching cables, and punching buttons to sculpt and shift the sound.[9]

Hackersports and Technological Hedonization

Hewitt's work can be thought of as a type of hacker sport, like a creative and funnier cousin to a nerdy hackersport like lockpicking. A mainstay at several hacker conferences like DEF CON, Chaos Communication Congress, and Hackers on Planet Earth (HOPE). The activity is often referred to as "locksport" to differentiate it from conventional lockpicking. Events vary widely in format, but all involve contestants picking locks in various competitive ways. For example, competitions can involve timed attempts to pick the same lock, chained together padlocks that need to be sequentially opened, or the "Locksport Wizard" format, where each contestant is given a sack of identical locks that they need to blindly pick (figure 2.3.5).[10] Typically, small groups of spectators and peers watch the competition in progress as contenders earnestly fidget with their tools and locks. As with most human endeavors, people have managed to "hedonize" work-related tasks like picking locks or constructing electronic circuits into playful activities.

In Hewitt's project, electronic prototyping components that might traditionally be thought of as the work-oriented tools of an electrical engineer have shifted into a playful application. History of technology scholar Rachel

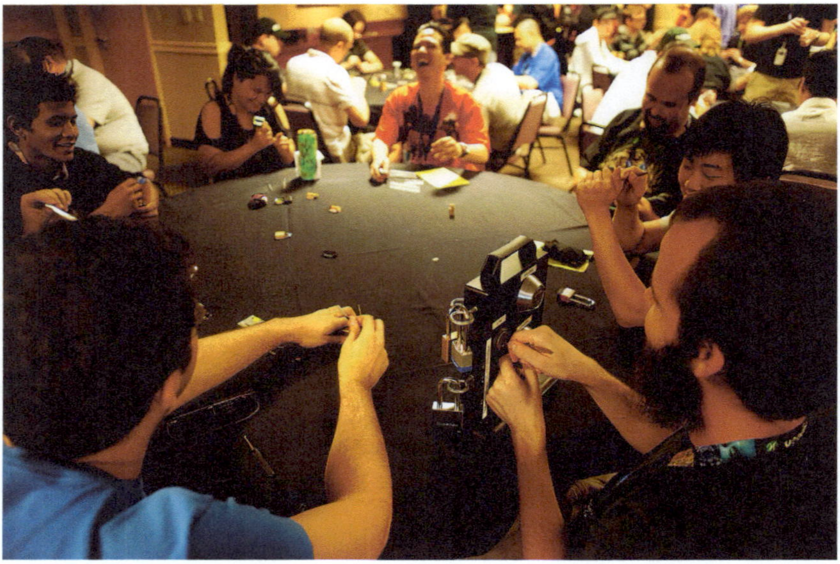

Figure 2.3.5
A lockpicking contest at the DEF CON 18 hacker conference in 2010. Courtesy Dave Bullock.

Maines would say the process has been "hedonized."[11] In her research, Maines tracks how work-oriented chores and their associated technologies have shifted into leisurely tasks and objects over time. Examples include gardening, hunting, cooking, needlework, home mechanics, and brewing. Even very gritty work-oriented tasks from the past, like coal mining or operating a steam engine, can shift over time from a task only done for labor into an afterlife of leisure, and this kind of use is often nostalgic of a past era. A lack of material hardship assists in do-it-yourself tasks becoming enjoyable hobbies: in times of prosperity and leisure, chores shift into artisanal crafts. The upswing of hedonization could be partially responsible for the "zombie media" effect of old technologies being reborn in mutated form.[12] Maines sees this type of playful practice as part of a larger cultural phenomenon, emerging when the utility or the seriousness of an object or practice is overshadowed by the pleasure of working with it. As a result, the hedonization of work-oriented tasks and technologies into pleasure-oriented activities forms an important core of Hewitt's work and DIY culture in general.

The Thrill of Impending Failure

The common ground that hardware practitioners like Darsha Hewitt, Bamanya Brian, Loud Objects, John Bowers, Tetsuo Kogawa, and many others share is a use of raw, low-level hardware to build audio from scratch. If commercial consumer electronic products are thought of as black boxes, this style of handmade electronic music starts without any boxes at all; the work is more about the playful construction of boxes, where the process of technical box building is transformed into ludic sport.[13] Writers like Derek Holzer go further to link this "hands dirty" electronic art to a renewed interest in the pre-digital pioneers from the 1960s and 1970s that built their own electronic instruments, including Nam June Paik, Steina and Woody Vasulka, Don Buchla, Serge Tcherepnin, Dan Sandin, and David Tudor.[14]

This desire to get dirty in the low-level wires and elemental circuits of electronic sound can be seen as a countercurrent to prefabricated options for creating music with contemporary software tools. As Holzer clearly explains, modern electronic music composition tools like Ableton Live enable beginners to construct electronic music with relatively few risks through their abundant presets, templates, and sample bundles. Ableton bills the software as "everything you need to make any kind of music."[15] While software-based approaches are useful for easily constructing electronic music, they offer very little in terms of an interesting live musical performance. Advanced software tools enable a performer to simply hit "start" and then blankly stare at their laptop and pretend to fiddle with adjustments for the remainder of a performance.

The live building of basic circuits works the other extreme of prefabricated audio. In this work, there is typically not even a score—the hardware schematic *is* the score. The tradeoff compared to a polished piece of music with little performative risk is a nearly unmusical piece of audio with immense performative risk. The risk involved with assembling low-level hardware is complete and catastrophic circuit failure, not just a misplaced note or something being out of tune. For Holzer, this high risk of the circuit performer's total failure is fundamental to the genre's attractiveness as an experimental art form: "For any art work to be considered experimental, the possibility of failure must be built into its process."[16]

Performative Conclusions: Learning from *20 Oscillators in 20 Minutes*

Instead of using a blackboxed piece of consumer electronics to generate audio, artists like Darsha Hewitt work to build audio-generating circuits from scratch, following a schematic instead of a score to guide their compositions.[17] This low-level construction of audio generators out of resistors, capacitors, chips, and wires can be seen as the antithesis to an electronic music performer staring at their laptop as it runs a prefabricated sequence. Instead of a polished piece of safely executed audio, Hewitt's piece is more a high-risk hacker's sport carried out through electronics and audio. This circuit-oriented work highlighted by Hewitt and others might be thought of as being "anti-prefab," and it playfully blurs the region between hardware hacker, music composer, and electronics educator.

The performance resembles many things: an old-fashioned telephone operator patching together phone calls, a live stand-up comedy routine, a competitive sport, and an educational demonstration in electronic circuitry. Hewitt calls it "a test of my technical abilities and an experiment in working with live troubleshooting as a method of musical improvisation"[18]—and in using only a schematic as a starting point, Hewitt follows John Cage or Fluxus in embracing the chance for errors in an indeterminate happening. In conclusion, *20 Oscillators in 20 Minutes* disposes of the standard black box of technology and replaces it with an imperfect performance . . . but this is exactly the point.

Theme 3 Building Identities

Identity and DIY Electronics: A Thematic Overview

Amy Spencer, when writing about UK and US zine culture, describes DIY practice as something an individual does to carve out their own identity.[1] Individuals often find themselves not reflected in different aspects of mainstream culture, and creative DIY production functions as an avenue for articulating and exploring the highly personal. In other words, the DIY mindset is often built as an extension of one's identity or to address something lacking in culture at large. Whether in music, haircuts, documentary films, print publishing, or clothing, DIY is regularly undertaken for personal reasons instead of with the aim of reaching a larger, universal goal. This also applies to the field of creative electronic technologies; artists, designers, and makers often approach the task of building or customizing technologies as an opportunity to bring personal issues forward. It could even be said that all the examples in this book are personal in some way. However, artists in this category pursue a more specific route when they use their cultural identities to bring DIY together with the body as political commentary.

Is All DIY "Personal"?

As a basic definition, Merriam-Webster defines *personal* as "of, relating to, or affecting a particular person: private, individual." As a result, the personal is infinitely complex and as variegated as humankind. However, this section of the book looks deeper at three different aspects of the personal in their own chapters. This cluster of case studies includes chapter 3.1, "Identity and *Taratter MI-03*: Parallel Worlds, Device Art, Chindogu, and Alternative Presents," which looks at the work of Maywa Denki as a personalized, worldbuilding endeavor. In this example, DIY-style techniques are used to build out custom devices that appear to have come from a retrofuturistic parallel universe. This worldbuilding can exist in one device or can be spread out over a collection of built things, and the world often reflects personal values, aspirations, or fears. As a result, the parallel universes that are fleshed out through these devices can ask why the world is the way it currently is, or they can act as an inspiration or cautionary tale about the future. In this chapter, Maywa Denki's identity is explored as a small-scale shop or store that builds whimsical technological devices. One particular device is the *Taratter MI-03*, a wearable contraption that straps onto a person's feet and hands, enabling them to knock mallets on the floor by tapping their fingers. On its own, this device might not be significant, but taken within a wider context, it brings up many important points, including how functionality in design is a shifting concept, how analog and digital systems can be linked in novel ways, and how intentionally awkward designs—also labeled in other contexts as "chindogu"—can be used to cleverly question why devices in the built world are the way they are.

The "personal" applies to wearable devices, clothing, and things used to adorn the body. For a significant portion of Western thought, the body has been envisioned as being separate from the mind—which undoubtably is wrapped up in René Descartes's *Discourse on the Method* from 1637 where he states "I think, therefore I am." This dualism between body and mind often portrays rational faculties as separate from the body and in control of it; the mind controls and the body follows directions. Part of the problem with this framing is that, among other things, women can be portrayed as more biological than men. In other words, according to Elizabeth Grosz from Monash University, "Women are somehow more biological, more corporeal, and more natural than men"—they are so enmeshed in feminine biology that being rational is difficult or problematic.[2] Similarly, lower-class individuals and people of color were also regularly portrayed as being more corporeal and less intelligent.[3] As a result, the "personal" is closely intertwined with discussions of gender, intelligence, and race—plus the topic of how subjectivity, corporeality, and identity are intertwined with wearable things.

Chapter 3.2 focuses on exploring the work of the Barbie Liberation Organization ("B.L.O."), a nonprofit collective of parents, artists, and hackers that launched one of the most innovative publicity hacks of the decade in 1993.

The project consisted of swapping the electronic voice boxes of two toys with each other: a Teen Talk Barbie who said things like "Math class is tough" and "Let's go shopping!" with a talking G.I. Joe doll that gruffly barked out "Vengeance is mine!," "Attack!," and "Eat lead, Cobra!" The B.L.O. swapped voices, modified the original packaging with leaflets and materials, then "shopdropped" them back into the original stores for other consumers to discover. The end result was a spectacular and historically significant example of culture jamming through DIY electronics. This is explored in the chapter "Identity and the *Barbie Liberation Organization*: Culture Jamming, Détournement, and Mediagenics."

In chapter 3.3, "Identity and the *Stock Market Skirt*: Gender, Telerobotics, and Clothing as Conversation," the robotic work of Nancy Paterson is used to show how DIY approaches excel at exploring the topic of gender. Paterson, a pioneering artist in the creative interactive media and network technologies, built a project in 1998 titled the *Stock Market Skirt* that was the first Internet-controlled telerobotic device to visualize data. In some ways, this was one of the first examples of the "Internet of Things." *Stock Market Skirt* is a custom-built dress equipped with an Internet-controlled motorized mechanism that raises or lowers the hemline based on live stock quotes: the higher the stocks rise, the higher the dress gets mechanically hiked up. More important than the technical breakthrough of rendering Internet data in a physical form, the project skillfully highlights the complex topic of gender within the domain of Internet technology and robotics. Technology is never *just* technology, it is always intricately intertwined with people, objects, bodies, places, practices, cultural values, and beliefs. Paterson reminds us that embodiment, gender, power, and control are vital topics to pay attention to when working with new technologies.

As people, the things we make on our own are personal in several dimensions—the concepts, approaches, or materials are all reflections of the person that made it. DIY practices in general are highly personal. Taken together, the chapters that explore personal aspects of DIY see them as at least three things: a path to explore parallel worlds, a creative way to comically subvert industry, and a way to embody the issues intertwined with gender that regularly get excluded from discussions about the development, use, and impacts of technology.

3.1 Identity and *Taratter MI-03*: Device Art, Chindogu, and Alternative Presents

The Parallel World Electronics Shop of Maywa Denki

Maywa Denki was founded as an "art unit" collaboration between brothers Masamichi Tosa and Nobumichi Tosa in 1993. The group operates as a two-person "parallel world electronics shop" and embraces a pseudocorporate identity, donning Japanese blue-collar uniforms reminiscent of the small and medium-sized manufacturing companies that helped launch Japan's economic growth after World War II.[1] Maywa Denki literally means "Maywa Electric Company," taking its name directly from the defunct business that the Tosas' father owned in Ako City, which manufactured electronic components for Toshiba and Matsushita between 1969 and 1979.

The parallel world that the Tosas set out to create through Maywa Denki is a conglomeration of artistic device exhibition, creative musical performance robots, and experimental product development. Maywa Denki describes their work as building "nonsense machines," quirky and highly customized electromechanical devices that are best known for whimsically outputting analog sound. Their first set of devices, released in 1993 under the banner of the *Tsukuba Series*, are elaborate electromechanical machine-music devices that use motors, solenoids, and percussion instead of speakers to produce audio. There are a total of thirty-five inventions in the series, and they all operate on the basic principle of converting electrical power into physical, percussive movement (figure 3.1.1). These electrically powered mechanical instruments take a basic electrical solenoid and multiply it into an absurdly detailed universe of odd instruments. One called *Guitar-la* features a set of six acoustic guitars splayed like a peacock's tail that are triggered with a pedal organ (figure 3.1.1, center). Another device that works like a programmable drum machine is called the *Koi-beat*, and is worn like

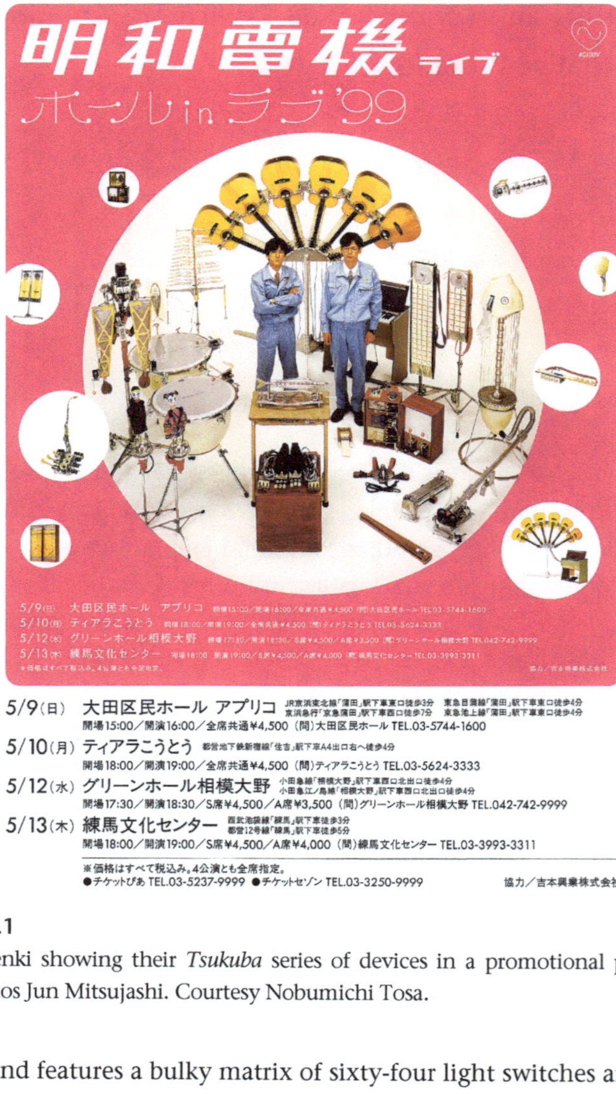

Figure 3.1.1

Maywa Denki showing their *Tsukuba* series of devices in a promotional poster from 1999. Photos Jun Mitsujashi. Courtesy Nobumichi Tosa.

a guitar and features a bulky matrix of sixty-four light switches and a turntable (figure 3.1.2, right).[2] These are odd devices that thoroughly have fun with their whimsical affordances and imaginative interfaces. The playfulness of the electrical contraptions contrasts with the refined mode they are presented in. They are rendered in highly refined materials, crafted with delicate care, and presented by official-looking individuals wearing powder-blue, shirt-and-tie uniforms. The one-of-a-kind whimsy of the devices is presented in a corporate-engineering style of deadpan.

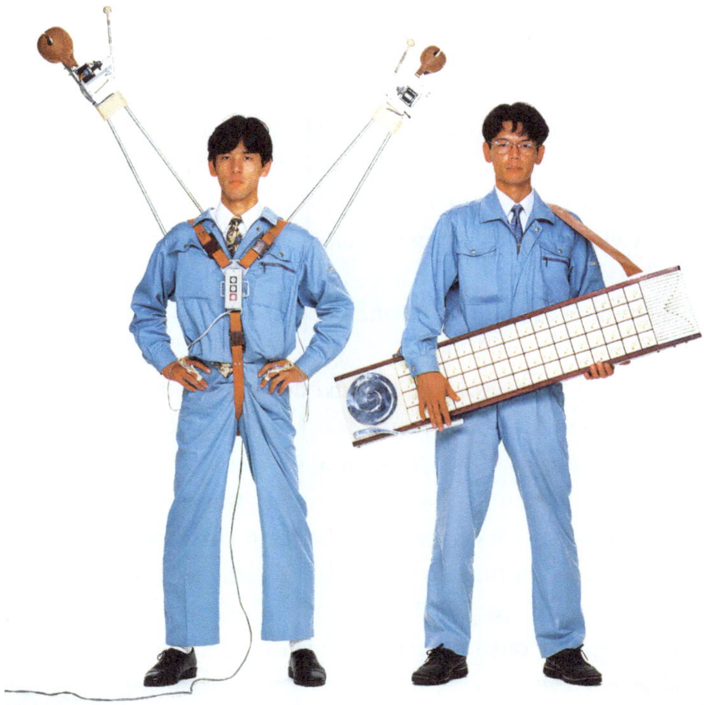

Figure 3.1.2
The Tosa brothers form the core of Maywa Denki, with Nobumichi Tosa (left) and Masamichi Tosa (right). Nobumichi Tosa shows the *Pachi-Moku,* a backpack-type of winged *Tsukuba Series* musical device operated by snapping one's fingers. Masamichi Tosa shows the *Koi-Beat,* a manual rhythm-making *Tsukuba Series* machine in the shape of a fish. Like *Taratter MI-03,* these devices work with relatively simple electronics and physical percussion. Photos Jun Mitsujashi. Courtesy Nobumichi Tosa.

Maywa Denki's work is more materially and technically refined than several of the other examples presented in this book. Although the work is independently built by the Tosa brothers in a relatively noncommercial setting, the work is presented as if it comes from a large corporation, and it borrows from industrial professionalism to pull this off. Corporate role-play is embraced by several other small-scale producers, including Survival Research Laboratories (chapter 3), the Barbie Liberation Organization (chapter 1.1), and the Institute for Applied Autonomy (theme 5). Among this group, Maywa Denki notably takes this role-play as its core identity—so much so that it is an actual company that wholly embraces its quirky, "parallel universe" persona. This does not negate its value as DIY practice that we can learn from. Rather, it serves as a reminder that a trend in DIY practice is that work straddles amateurism and professionalism, and that categories of DIY electronic art often overlap.

Personal Technologies and the *Taratter MI-03*

A useful case study of one of the *Tsukuba* devices is *Taratter MI-03*, which serves as a clear example of a personal form of do-it-yourself technology: it is highly customized, playful, and a piece of wearable electronic technology that amplifies the human body. Looking at this device in some depth provides clues about the leverage that small and personal technology production has in larger culture. In particular, *Taratter MI-03* shows that DIY approaches have the luxury of being able to pursue alternate visions of future technologies far outside everyday norms. These visions provide deep cultural value by playfully helping society reevaluate the social and technological norms in culture at large.

Maywa Denki describe *Taratter MI-03* as remote-controlled tap dance shoes that are designed to enable anyone to tap easily (figure 3.1.3). They explain that the "tapping sound is made by controlling the knocker set on the toe with fingers. Comfortable vibration cures foot problems."[3] In other words, tapping with your fingers will activate percussive mallets beneath your feet—with the tongue-in-cheek health benefits of massage. From a physical perspective, *Taratter MI-03* consists of two hand components and two foot components (figure 3.1.4). The hand components each have a flat aluminum plate that runs the entire length of the forearm, with three paddle-like keyboard buttons underneath the fingertips. Blocky platforms are also attached to the underside of the performer's shoes with three finger-like percussive knockers extending from each like electromechanical toes in front of the performer's feet. The hand keyboard gauntlets strap to the arms with tan leather straps, and each arm piece has wires that electronically connect to its corresponding foot actuator. When the performer presses the finger pads, the corresponding mallet immediately hammers the floor, producing a wooden-toned knock.

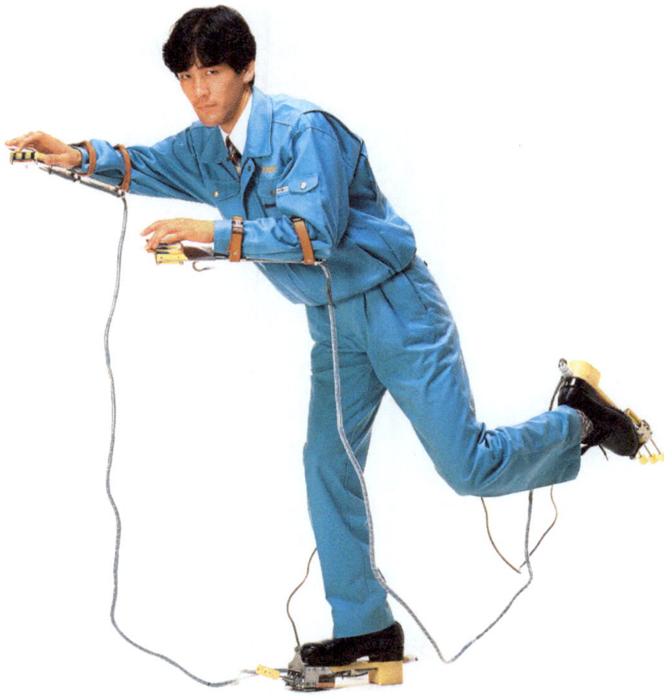

Figure 3.1.3
Nobumichi Tosa demonstrates the *Taratter MI-03* system by Maywa Denki (1993). 390×100×170mm (foot) 410×75×40mm (hand) / 2.3kg / AC100V 50/60Hz. Photo Jun Mitsujashi, courtesy Nobumichi Tosa.

The system technically works by finger switches turning on solenoid actuators in the feet; this electricity-into-movement theme underpins the entire *Tsukuba Series* of devices. The physical construction is materially refined, with muted and matching colors. *Taratter's* fit and finish is superb, making it look like a professionally engineered product despite its humble two-person development team. This stands in stark contrast to more visually cacophonous works like Nam June Paik's *Robot K-456* or Survival Research Lab's *Demanufacturing Machine*.

In order to have a better understanding of how this style of DIY electronics work has cultural value, two terms that are useful to explore in more depth: "chindogu" and "device art." Together, these concepts contribute to a vocabulary around DIY practices and its process of challenging social and technological norms.

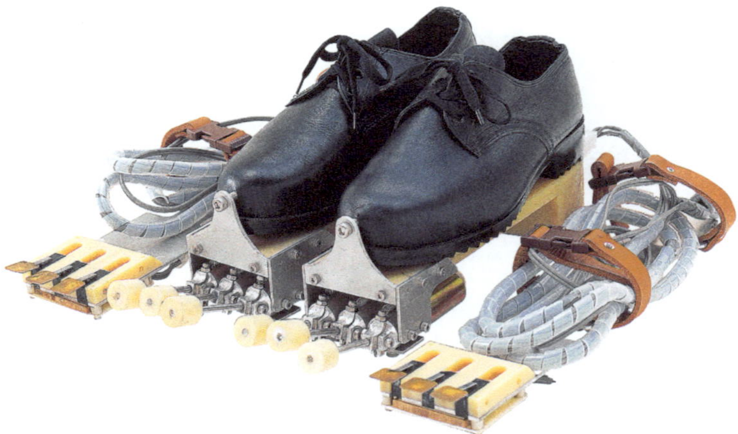

Figure 3.1.4
Detail of the *Taratter MI-03* system by Maywa Denki. The system is worn on the hands and feet, with finger movements triggering the foot-worn actuators to percussively tap the floor. Photo Jun Mitsujashi, Courtesy Nobumichi Tosa.

The Parallel World of Chindogu

Taratter MI-03 resembles an oddball invention, a bit like an outlandish Rube Goldberg drumming device. To be fair, Rube Goldberg only drew cartoons of odd devices and did not build them, but Goldberg and the Tosas both clearly take more pleasure in whimsy than in efficiently solving a practical problem. Steven M. Johnson is a similar illustrator of amusing inventions, and he has been illustrating speculative devices since the mid-1970s. Johnson's work includes inventions like hot tub automobiles, cigarette-smoking helmets with integrated air purifiers, and "fear furniture," including a bulletproof shower and an earthquake-secure dining canopy for your living room.[4]

For those well-versed in experimental design, the work of Japanese designer and humorist Kenji Kawakami and his genre of "chindogu" physical inventions may come to mind. Chindogu, literally translated as "odd tool" or "distorted tool," refers to whimsical inventions that humorously solve one practical problem but create more problems than they solve. Unlike Goldberg or Johnson, Kawakami's devices are physically built and photographed. Examples include a bath body suit, a walk-and-wash ankle-attachable laundry tank, and a mop-like onesie for a crawling infant to help wash floors.[5] Chindogu objects often parody inventors being blinded by a monocular focus on problem solving and losing sight of the larger social context. In the case of the *Baby Mops* device, the humor comes from the sacrilege of a parent putting a baby to work by dressing them in a wet, dirty mop. It cleans your floors at the expense of having child protection services take your child away, so to speak.

Although Kawakami's chindogu and the Tosas' Maywa Denki both originated in Japan in the early 1990s, they differ in that chindogu is overtly a joke *on* industrial design. Chindogu is an intentional misreading of social cues; it plays with and flaunts social norms through awkward design. Maywa Denki, on the other hand, is more a playing *with* industrial design.[6] In other words, Kawakami's work is intentionally "bad" industrial design, while the Tosas' work is more aptly categorized as "alternative universe" industrial design.

In terms of common ground, both clearly radiate the joy of inventing and building things. In addition, *Baby Mops* and a device like *Taratter MI-03* share that they clearly embrace an alternate set of design values from most standard consumer products. They are both mischievous and unconventional—even fantastic. However, chindogu and Maywa Denki have significant differences in terms of the parallel universes they embrace: chindogu's universe exists in intricately executed gags, while Maywa Denki is focused on exploring the elegance of a single mechanism or metaphor.

The cultural value we can extract out of the parallel worlds of chindogu and Maywa Denki includes the same valuable things we can generally draw from other forms of personal DIY production. Chindogu's core contribution is that it highlights the unintentional chindogu that fills the world around us—the high-technology devices that solve one problem well but cause a host of other unintended negative consequences. Google Glass serves as a straightforward example of an unintentional chindogu: a technologically advanced product that grossly oversimplifies culture, society, and what it means to interact with other people.[7] Joi Ito might classify the problem with chindogu as being the same problem that flows out of Silicon Valley—technologists thinking that the world's complex problems can be solved through technology alone.[8] Instead, the world is much more complex.

On the other hand, playing with the complexity of the world is the core of Maywa Denki's cultural value. Maywa Denki—like other DIY electronic and mechanical projects—drills down into a personal, obsessive pursuit of a thing. In the case of the *Tsukuba Series* of works and *Taratter*, the "thing" is an electromechanical solenoid. Maywa Denki builds out a personalized "solenoid-world" through their series of devices. This deep and unusual dive into a single tool, technology, or metaphor is notable as a general characteristic of the DIY mindset; do-it-yourself and independent producers often have the luxury to focus on a singular fetish that appears irrational to most.

The payoff of following an irrational fetish is raw innovation—although the practicality of the pursuit is often questionable. In the case of Maywa Denki, the built device is so odd that it renders mass-produced consumer products significantly bland. This hits the core of what makes many other independent and DIY projects valuable: by tilting the horizon of reason through the incessant pursuit of building an object, the sphere of possibility of what humans can produce is innovatively opened up.

Contexts in Japanese Device Art

In addition to understanding chindogu, the concept of "device art" is also useful for providing a context for Maywa Denki's work and the personalized approach of building artistic electromechanical machines. Machiko Kusahara introduced the term "device art" in 2006 to describe a significant trajectory in Japanese art where technology is the essential core of the work, playfully combining art, design, entertainment, popular culture, and commercial products.[9] Maywa Denki exemplifies this concept, where unusual devices circulate as artistic musical instruments, but are later reengineered into commercial products. In the case of the *Taratter MI-03* and the *Tsukuba Series*, Maywa Denki repurposed their solenoid-music concept into a set of inexpensive, simplified toys under the banner "Knock." Kusahara's first argument with the term device art is to argue that the Western lines of demarcation between fine art and commercial product design do not exist in the same way in Japan. In many instances, DIY electronic art can be clearly thought of as "device art."

Another attribute that Kusahara assigns to device art is a playful and whimsical approach to electromechanical work with technology, which she sees as originating out of the larger cultural impact of the industrial revolution in Japan. In the UK, Europe, and North America, the Industrial

Revolution (1760—1840) brought technological changes such as steam power, mechanical tools, textiles, agricultural advancements, and new forms of transportation. However, it also produced the factory system, pollution, and a population increase that led to dehumanizing work conditions. As a result, a significant counterindustrial sentiment emerged as a response, from Luddites smashing automated looms to film works like Charlie Chaplin's *Modern Times* depicting the negative components of mechanization and industrialization. Compared to the West, Japan's industrial revolution happened almost a full century later—starting around 1870 instead of 1760—and was rolled out as a state-wide initiative (the "Meiji Restoration") to have Japan catch up technologically to the West. Kusahara sees that the later introduction of industrial technology resulted in a more positivist and affirmative acceptance of industrial systems and automation. This recognition is due in large part to Japan's long period of peace before industrialization, and as a result, new technologies and scientific instruments took hold as incredible and fascinating cultural phenomena embraced widely throughout Japan. "Technology was something to be enjoyed rather than something to be feared."[10] Put differently, Kusahara's argument could be related to more mature industrial technologies being introduced to Japan than in Britain, for example, with more of the technical and social difficulties already worked out.

The economic and technological history of Japan and larger cultural attitudes toward new technology are outside the focus of this text. Although Japan does not have a monopoly on playful, optimistic attitudes toward technology, it is worth acknowledging the country's unique social variables. At least on the surface it seems true that Japanese artists seem to actively and creatively bend the boundaries between fine art, industrial design, and commercial products. However, Western-style technological optimism in art is as old as at least Experiments in Art and Technology in the 1960s, and has been a strong force behind work under the banner of "new media" at festivals like Ars Electronica, SIGGRAPH, Future Everything, ISEA, and others. Similarly, a kind of "deviceness" also exists in pockets like the San Francisco Bay Area and in institutions like the MIT Media Lab or the NYU Interactive Telecommunications Program. Device art is perhaps more useful as a category of contraption-oriented work that has emerged from a host of variables in a number of different locations. It is valid, however, to claim that Japan is a unique leader in this style of work. Device art is also acutely useful in highlighting important work that mashes up the domains of fine art, design, technology, and entertainment.[11]

The Speculative Futures and Alternative Presents of DIY Practices

In particular, *Taratter MI-03* and the rest of the Tsukuba series are driven by a desire to reanimate musical techniques from the past, specifically non-digital forms. These are nonsense in the way they reject the contemporary logic of digital music, Spotify, Tidal, or sampled and remixed music. The *Tsukuba series* breaks the logic (or "sense") of the present mode of doing things and strips down music into a simple, percussive hammering. In this way, *Taratter MI-03* is simple, elegant, and self-contained, and is physically embodied and substantial instead of abstract and purely digital.

This sidestepping of standard history is a method commonly used by science fiction writers to create engaging scenarios: for example, what would the current world look like with different world war outcomes? These alternate or alternative presents are akin to a hypothetical situation constructed as if history would have continued down an alternate path. Design theorist James Auger envisions that these alternative presents can be visualized as a side-stepped pathway for product development (figure 3.1.5).[12] Auger sees that this alternative approach is useful because it helps challenge and question the objects and systems that are produced by current modes of product design and manufacturing. In practical terms, Maywa Denki uses outdated percussive technology from the past that is taken and put into the center of the present, as if history would have continued down an alternate path. This is core to Maywa Denki's parallel universe that embraces a type of obsolete technology as a winner instead of the contemporary logic of seeing it as an outmoded loser. Designers like Kaiwen Yang have expanded on Auger's diagram to envision that alternate histories can function to bend the present-day into a more preferable state (figure 3.1.6).[13] In other words, Yang argues that by fabricating the speculative, we actively bend reality into a parallel universe.

Conclusions: The Alternative Present of Archaeological Neoretroism

In conclusion, *Taratter MI-03* is significantly more than a piece of slapstick industrial design. It is a far cry from the "chindogu-esque" slapstick

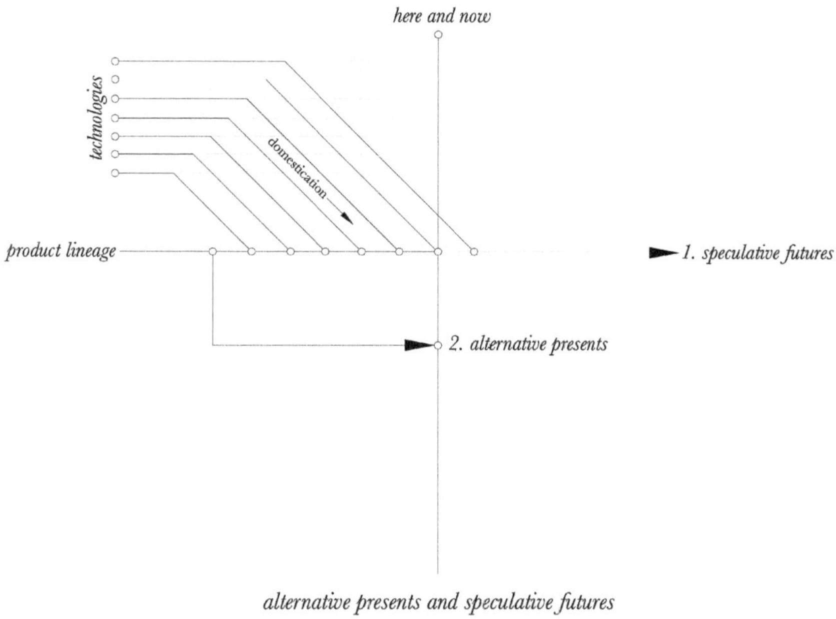

Figure 3.1.5
James Auger (2010). In this diagram, Auger illustrates how industrial design work can model out speculative futures by manipulating product design lineages. Courtesy James Auger.

industrial design like Simone Giertz's "shitty robots," where devices are built to be painful or humorous failures.[14] An example from Giertz is a robotic arm built to apply lipstick, but when functioning, it indiscriminately and forcefully slathers lipstick in the general area of the mouth.[15] Instead of engaging in slapstick physical comedy like chindogu, *Taratter MI-03* offers considerably more: it digs into a particular technique more like a form of media archaeology—the old-fashioned or backward technique of using mechanical movement to produce analog audio—and builds up a type of parallel world of devices in the process. Maywa Denki refers to these as "nonsense machines," but not in a derogatory way—this is more a shunning of the mundane and the mass produced.[16] "Wonder machines" might be a more accurate label.

In this way, personal DIY electronic objects and device art show us that playfulness often does not equal a headfirst dive into new technologies. These projects are often playful and humorous because they are

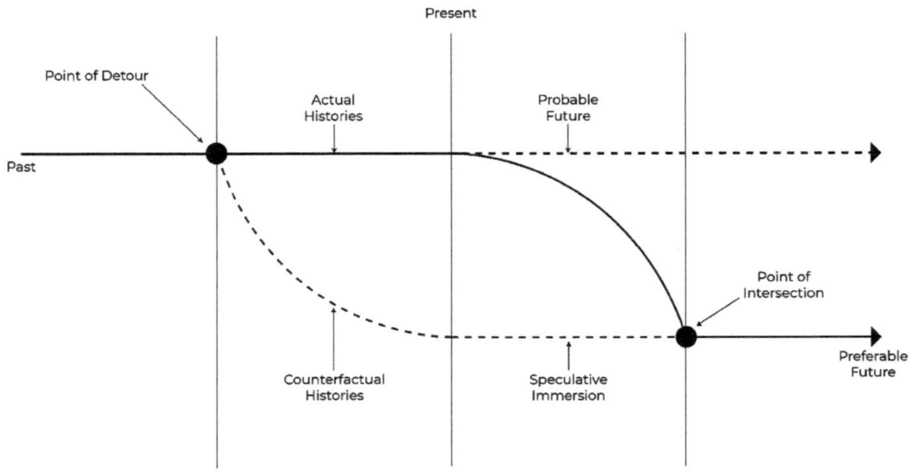

Figure 3.1.6
Kaiwen Yang's diagram of how parallel worlds and counterfactual histories can "bend" history from its probable path into a preferable path. Courtesy Kaiwen Yang.

technologically "backward" in some way, mixing mechanical interfaces, movements, or the old-fashioned with the present. This is a backwardness, or an archaeological "neoretroism." The present is reinterpreted through the lens of the reanimated past, a zombie detour across the history of technology. Resurrecting history is significantly more than nostalgia in this case, however. It is a neoretro style of media archaeological studio practice that helps us find the future in the past.

3.2 Identity and the *Barbie Liberation Organization*: Culture Jamming, Technical Détournement, and Mediagenics

In July 1992, Mattel released an electronic Barbie that said, "Math class is tough," causing a significant and organized backlash by the National Council of Teachers of Mathematics and Science and the American Association of University Women.[1] These groups, along with members of the public, opposed the phrase for reinforcing the gender stereotype that women are innately less capable at math than men. Initially, Mattel offered a free exchange for dolls that did not include the phrase. However, after continued pressure, it permanently removed the phrase from the toy's electronic lexicon in October 1992.[2] Although Mattel removed the comment about math, vapid phrases remained. Barbie talked about shopping, the beach, clothing, and weddings.

The Birth of the Barbie Liberation Organization

Working out of the East Village of New York City, a cluster of artists, parents, and technologists radically modified the Teen Talk Barbie to say things that would shock consumers and garner media attention. Their plan was relatively simple: take a talking Barbie and a G. I. Joe and switch their electronic voice boxes (figure 3.2.1). The result of the circuit board exchange was like a brain and voice transplant—switching electronics swapped stereotyped voices into the opposite gender. Teen Talk Barbie snarled combative phrases such as "Eat lead, Cobra!" "Attack!" and "Vengeance is mine!" while G. I. Joe excitedly proclaimed, "Let's plan our dream wedding!" and "Let's go shopping!"[3] It was simple but brilliant.

They called their initiative the Barbie Liberation Organization (BLO). This group of anonymous activists was led by artist Igor Vamos (a.k.a. Mike Bonanno), who had recently run creative protests in Portland.[4] Vamos is now

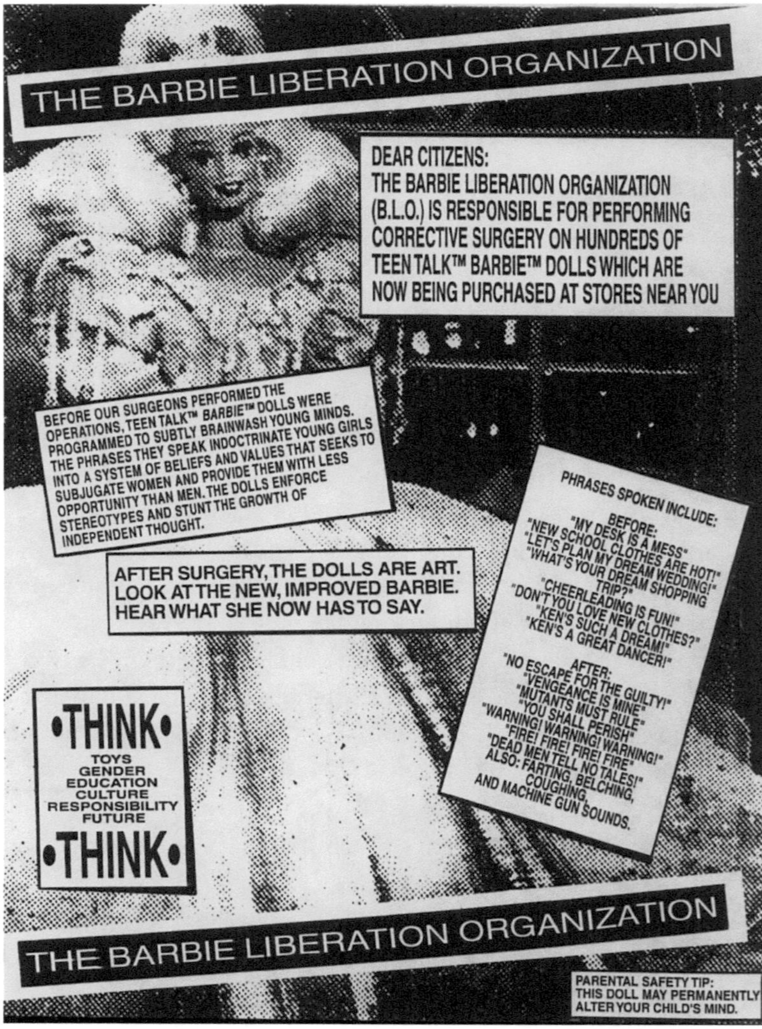

Figure 3.2.1
Barbie Liberation Organization leaflet, 1993. Courtesy The Brian Sutton-Smith Library and Archives of Play at the Strong Museum.

better known as one of the Yes Men—a political activist group that specializes in staging intricate hoaxes while impersonating corporate spokesmen.[5]

Culture Jamming Identity through News Media

This chapter explores the concept of culture jamming when applied to artistically oriented DIY electronic practice. Mark Dery, who published the influential *Culture Jamming: Hacking, Slashing and Sniping in the Empire of Signs* in 1999, defines culture jamming as "media hacking, neo-Situationist sociopolitical satire, and guerrilla semiotics, all in one."[6] Dery casts the net of practice for culture jamming to be quite wide, including media hoaxers, billboard bandits, pirate broadcasters, fan fiction "slash" authors, and other hijackers of official forms of corporate communication. Through modification, these messages are spun into politically subversive or obstinately personal ends.

In other words, culture jamming is the process of hacking communication media through things like billboard or magazine advertisement modification, in a form that flips their meaning from the original "please buy this product" message to a political message or social commentary on the brand, product, or company doing the original advertising.[7] The magazine *Adbusters*, for example, can be thought of as embracing this spirit. Hacking also works for things other than just images. It can work with technology, too. In other words, electronic hardware can be jammed just like visual advertising and communication media.

The BLO borrows part of its name and approach from the Billboard Liberation Front (BLF), a tactical art group founded in 1977 "fueled by a single passion: the timely improvement of outdoor advertising."[8] The BLF operates as a faux advertising agency that modifies publicly displayed outdoor billboards to subvert or critique corporate advertising campaigns. The BLF typically works by selecting an advertisement, preparing large format hard copy printouts to amend the billboard that match the color and typeface of the original ad, and pasting up their panels on top of the billboard image, usually at night.[9] The BLF documents its work through before-and-after photographs of billboards with an accompanying press release that sarcastically describes the campaign done in cooperation with the advertisers to improve their messaging. For example, in 2010, BLF modified a McDonald's Egg McMuffin billboard in San Francisco from saying "I'm lovin' it" to

"I'm sick of it," and "You have about 10,000 taste buds—use them all" to "You have about 10,000 taste buds—kill them all."[10] In other words, panels of "sick of" and "kill" were pasted on the billboard to subvert its original message.

The Barbie Liberation Organization and Technical Détournement

Meanwhile, the mission of BLO was to free Barbie from sexual stereotypes through creative hacking. Feminist technology scholars have long explored how gender is perpetuated through technology design. Feminist pioneer Judy Wacjman, in her classic text *Feminism Confronts Technology*, writes that "machinery is designed by men with men in mind," and for this reason, "industrial technology reflects male power as well as capitalist domination."[11] In other words, technological design also involves the reproduction of cultural understandings and even stereotypes. For this reason, Anne Balsamo refers to technology and culture combined as "technoculture."[12] Balsamo sees design as a social process that can either reproduce existing biases or open up space for reconfiguring cultural dynamics of gender. Design as a form of innovation implies "the creation of unique arrangements that provides the basis for a reorganization of the way things will be in the future."[13] Technology design, in particular, gels different biases into objects and infrastructures.

While discussing what could be done to address Teen Talk Barbie and its stereotypes, some members of the Barbie Liberation Organization raised concerns about stereotypes in other children's toys, including the war-oriented, male G. I. Joe made by Mattel's competitor Hasbro. After confirming that a talking "Electronic Battle Command Duke" G. I. Joe doll—which yelled out violent tidbits like "Dead men tell no tales!" and "You shall perish!"—was currently on store shelves, the group came up with a plan to swap the electronic voice circuits, inverting the "yin/yang figures of American gender construction."[14]

Working in the East Village of Manhattan, the group figured out how to swap the circuit boards inside the two toys using screwdrivers, a saw, a soldering iron, glue, and wire (figure 3.2.2). First, both toys had to be pried and cut apart and have their battery and speaker connections desoldered. Then the circuit boards could be swapped, resoldered, and reassembled. The task was not as simple as a straight swap, however. As a result of G. I. Joe's body cavity being significantly larger than Barbie's, his circuit needed to be

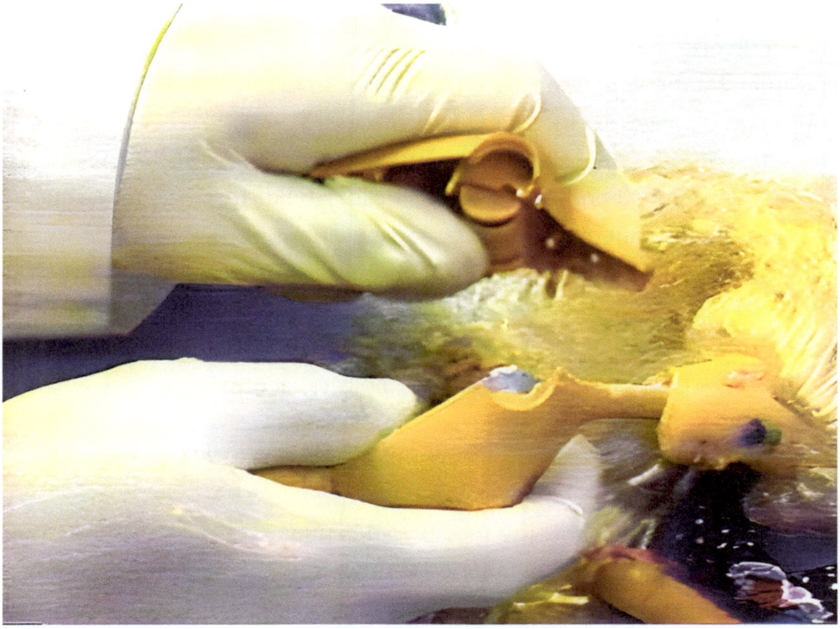

Figure 3.2.2
Barbie Liberation Organization: Video still of electronic circuit swap between G. I. Joe and Barbie. Courtesy Igor Vamos.

physically cut down, with approximately one-third of it removed in order to fit. Putting the dolls back together required bending capacitors, shaving down the interior of the toys, and using shims, a nail wrapped in duct tape, and plenty of hot glue and epoxy.

The circuit board swap transplanted the stereotyped voices into the opposite body. The result was a moment of humor that acts as a contradistinctive reevaluation of gender stereotypes and the worldview that produces them. By removing their electronic voice boxes, BLO reversed their roles, and in the process, worked to critique the construction of gender. Somewhere between a dozen to 3,000 toys were modified by the group, placed back into their original packaging along with printed materials (figure 3.2.1), and covertly returned to the shelves in a kind of reverse shoplifting (or "shopdropping") for consumers to purchase. The tactical DIY approach of the Barbie Liberation Front needed media attention, and they chose the Christmas shopping season of 1993 as their target.

Culture jamming can be seen as acquiring methodologies from a number of diverse sources, including *détournement*, a technique of appropriation outlined by the Situationists in the 1950s.[15] In détournement, well-known objects and images are used and taken through a *detour* to create an alternate message, often in oppositional contrast to the original source. The original is directly sampled and turned on its head: similar to satire, burlesque, or parody, détournement directly uses large portions of the original work and hijacks and derails it. The Barbie Liberation Organization accomplishes this through subversive, targeted hardware remixing. In a similar way, DIY electronic practices often work as a style of subversive détournement, using common objects as a starting point for political action.

The Mediagenics of Operation Newspeak: Christmas 1993

Operation Newspeak was the press-based campaign during the 1993 holiday season in support of BLO. The organization sent press kits with a VHS press release tape and a printed press release to journalists and news organizations around the United States. The VHS tape featured a static Barbie face photograph overlaid with a live action video mouth that narrated the mission statement of the group and gave a brief technical overview of the project (figure 3.2.6). The script was written from the perspective of a Barbie doll that desires liberation from her confined gender stereotypes. The BLO timed the release of this kit to reach news outlets on Christmas Eve 1993, with the intention that it would gain traction on Christmas Day after individuals called news outlets to report their electronically hacked dolls.

BLO also released a one-page handout titled "Official Barbie Liberation Organization Barbie/G. I. Joe Home Surgery Instructions" that detailed the toy modification in eighteen illustrated steps along with the required tools and supplies (figure 3.2.3). The DIY guide resembled a comic book-style layout, consisting of text in photocopy-friendly diagrams. The customers who unknowingly bought the modified toys had reactions including shock, bewilderment, humor, and delight. The modified doll packages included an assortment of leaflets (figure 3.2.4) with information about the project. A telephone number was also deployed—one that went to an answering machine message that started with "You've reached the phone number for the Barbie Liberation Organization. The BLO is dedicated to the humanization of children's toys . . ." The prerecorded message encouraged parents

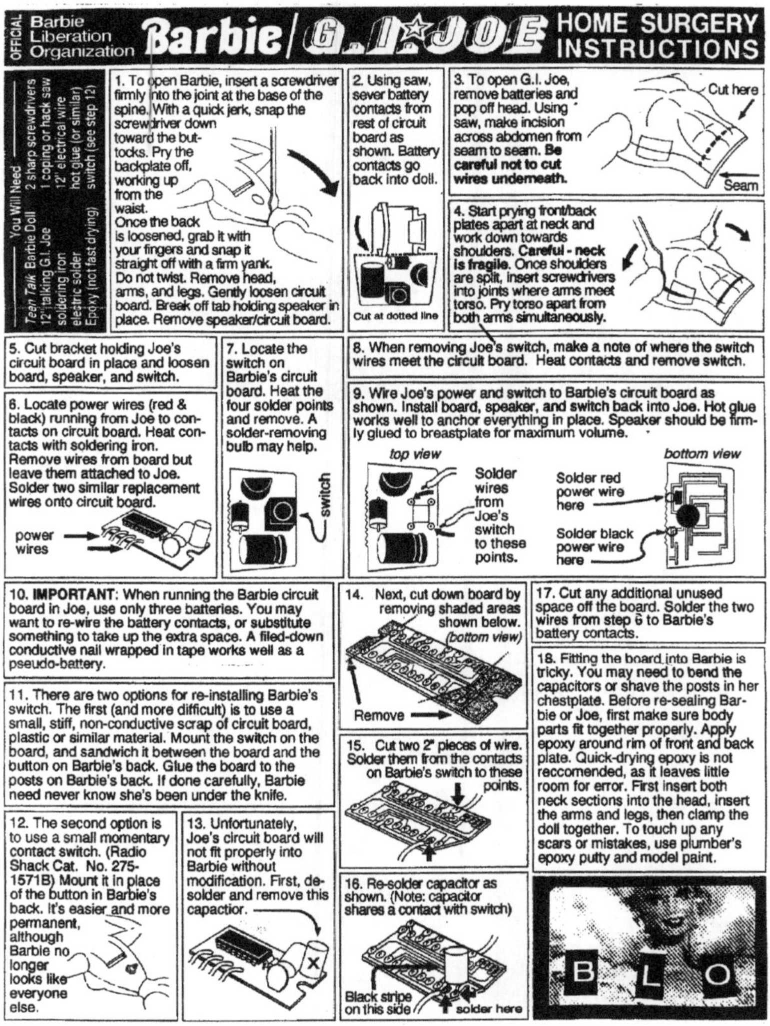

Figure 3.2.3

Official Barbie Liberation Organization Barbie / G. I. Joe Home Surgery Instructions that detail the toy modifications in eighteen illustrated steps along with required tools and supplies. Courtesy Igor Vamos.

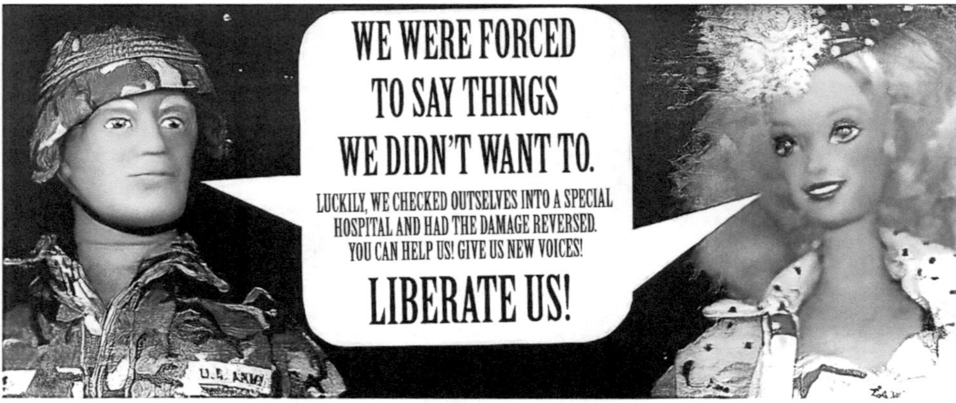

Figure 3.2.4
Barbie Liberation Organization promotional materials featuring a conversation between a G.I. Joe and a Barbie doll. Courtesy Igor Vamos.

to agree with the organization's anti-sexist and anti-violence stance, and to contact their local television stations and newspapers. The photocopied instructions to perform the hack and the invitation for the public to participate in activism invited the public to intervene in the commercial distribution of products and to question political and economic systems (figure 3.2.5).

It is worthwhile to note in this piece that the electronic component was only a part of the larger project. The "mediagenic" components—or the components of the project dedicated to communicating clearly to the press—made up a large part of the entire system.

Press Reaction and Mediagenics

The Barbie Liberation Organization went to significant lengths to gain media attention, including recruiting two children to talk to the press about the modified dolls.[16] The group also kept a number of modified dolls in reserve and planted them in stores for the media to find. When news organizations called the BLO phone number for more information, the group would figure out the nearest toy store to the caller and quickly scramble to shopdrop a modified G. I. Joe or Teen Talk Barbie there. According to Igor Vamos, "On at least one occasion, BLO members were still in the store when the journalist arrived. They watched him find the toy, test it, and

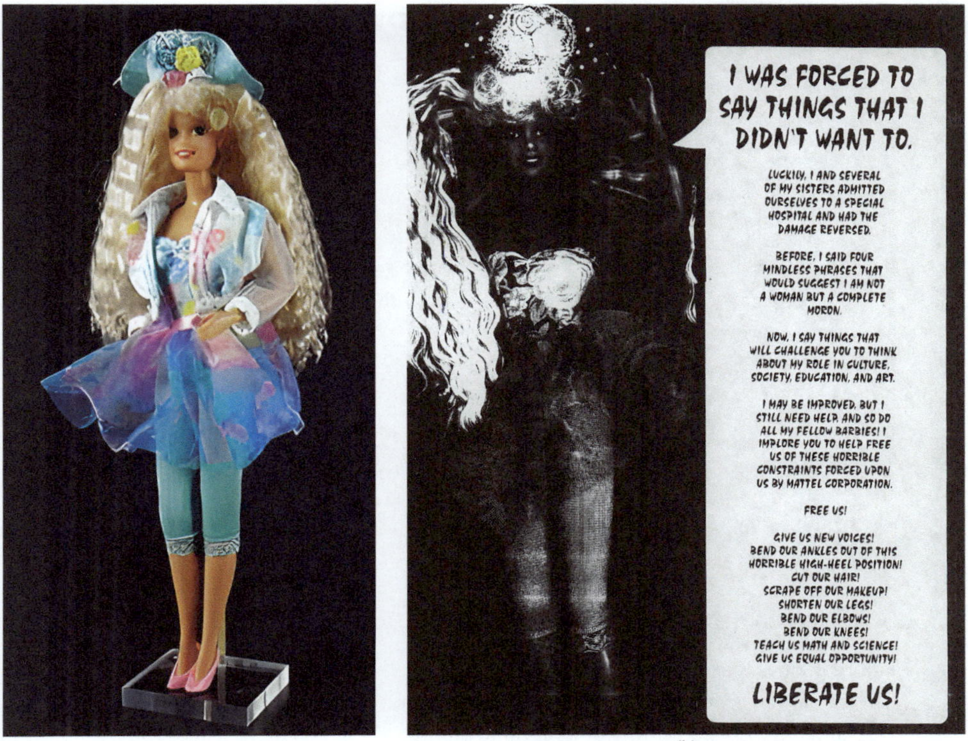

(a) (b)

Figure 3.2.5

The Barbie Liberation Organization initiative produced an array of printed materials that promoted the project and which were inserted into the modified toy product packaging. Courtesy The Brian Sutton-Smith Library and Archives of Play at the Strong Museum.

triumphantly purchase it—proof positive of the power and reach of the Barbie Liberation Organization."[17]

As an indication of the mediagenics of their operation, the BLO received widespread dispersion and attention, including an in-depth segment about the project on *Dateline NBC* using hidden identities (figure 3.2.6). One child that received a modified toy, Hannah Henzey, stated to a local CBS news crew in San Diego that she thought the toys were clever in how they shifted gender roles, saying, "Well, men can say they want to go shopping too."[18] In an interview with the *New York Times*, a BLO representative made the mediagenic leanings of the prank clear, stating that "our goal is to get media attention. We are trying to make a statement about the way toys can encourage

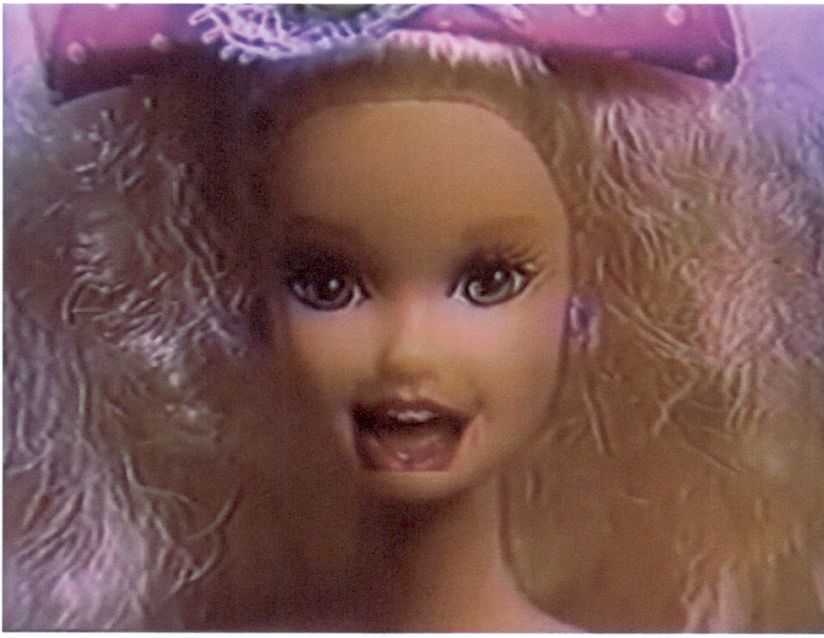

Figure 3.2.6
Screen capture from Barbie Liberation Organization press release VHS video, dated December 22, 1993. Courtesy of Igor Vamos.

negative behavior in children, particularly given rising acts of violence and sexism."[19] The success of BLO underscores the effectiveness of DIY tactical media. Mattel, the maker of Barbie, "was slightly amused if only because it knew it could produce [many] more dolls than could ever be modified."[20] However, when evaluating the immense national press that the BLO project received, it is hard not to interpret the company's response as a calculated public relations strategy to handle a troubling situation and stock price that was spiraling out of control.

Conclusions in DIY Electronics as Culture Jamming

The Barbie Liberation Organization culture jammed through the deceptively simple electronic DIY hack of swapping two voice boxes. The technically straightforward swap resulted in an innovative challenge to stereotyped gender roles in America, the country's preconceptions about intelligence and attraction to violence, and its mediagenic approach took advantage of the American news cycle.

If looking for a strong example of a successfully "jammed" DIY technology, BLO provides a premium one. Like Nancy Paterson's *Stock Market Skirt*, this project uses DIY approaches to reinsert the topic of gender back into the field of technology design. It worked because it leaned on the famous cultural icons of Barbie and G. I. Joe, and it worked with them in a way that was clearly legible to the public. Children, unexpected humor, careful preparation, and the inversion of famous cultural icons made this DIY intervention a mediagenic success. Toys often echo the values of wider culture, and as pieces of Americana, Barbie and G. I. Joe serve as useful platforms with which to criticize American culture.[21] At the core of it, however, was a simple electronic hack of swapping two toy voice boxes with each other. In electronics terms, this is elementary—but it is also absolutely transformative, surprising, and revealing when deployed in a mediagenic and photogenic format through the press.[22] The Barbie Liberation Organization is the author of one of the all-time most clever hacks to consumer products, ever. As such, it serves as an inspiring example of the DIY mindset to artists, hackers, designers, and others with an eye toward having a political impact on the world.

3.3 Identity and the *Stock Market Skirt*: Gender, Telerobotics, and Clothing as Conversation

Stock Market Skirt operates a bit like a haunted piece of futuristic clothing, where the hemline of a dress gently rises and falls with a soft, robotic murmuring of stepper motors. The project, built by Canadian artist Nancy Paterson in 1998, consists of a custom-tailored blue taffeta dress with a fitted black velvet bodice midsection and blue, cape-like butterfly sleeves. The dress sits on a beige dressmaker's mannequin that lacks a head, arms, and legs (figure 3.3.1). If careful attention is paid to the skirt's hemline, one notices that it physically twitches and moves slightly every few seconds. The hemline rises and falls, with the crisp blue taffeta gathering up like a rising and falling theater curtain (figure 3.3.2). Around the mannequin are several computer monitors displaying large white text on a blue background; upon closer inspection, these are recognizable as stock exchange symbols and prices. The screens scroll text from right to left across a blue background, visually resembling a live ticker display on a trading room floor. It also gradually becomes apparent that the higher the stock price fluctuates, the higher the hemline rises, as the dress dynamically echoes the stock exchange's trends. In other words, when the stock price goes up, so does the skirt.

The project, which initially launched at the Bell Centre for Creative Communications in Toronto in January 1998, is a pivotal work in understanding how technology, gender, and the body can be intertwined through DIY electronic art. Paterson's work reminds us that DIY approaches to technological production often explore personal themes, and that DIY production is a useful avenue to communicate marginalized experiences and perspectives about gender.

Nancy Paterson (1957–2018) was a Toronto-based artist who was pivotal in infusing the field of media art with questions of gender. Over the course

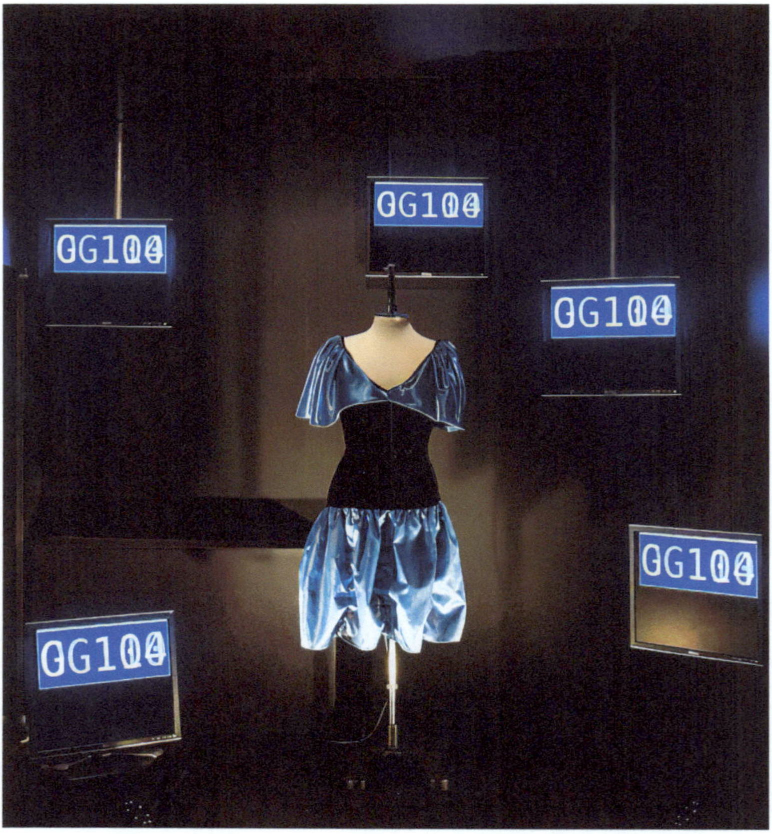

Figure 3.3.1
Nancy Paterson's *Stock Market Skirt*, 1998. Photo courtesy of Scott Massey.

of three decades, she created technically complex and socially engaged work. Her early media-based artworks incorporated images and devices from domestic environments in the 1950s, regularly depicting women interacting with space-age domestic technologies. By 1982, Paterson was one of the first Canadians on the public internet, connecting through the University of Toronto Zoology Department's "utzoo" network. Utzoo introduced Paterson to internet email through a Unix command line on a telnet connection, operated by Henry Spencer, who also ran the first active Usenet site outside the United States, starting in 1981.[1] *Stock Market Skirt* was Paterson's first work connected to the Internet; up until that point, her media-based

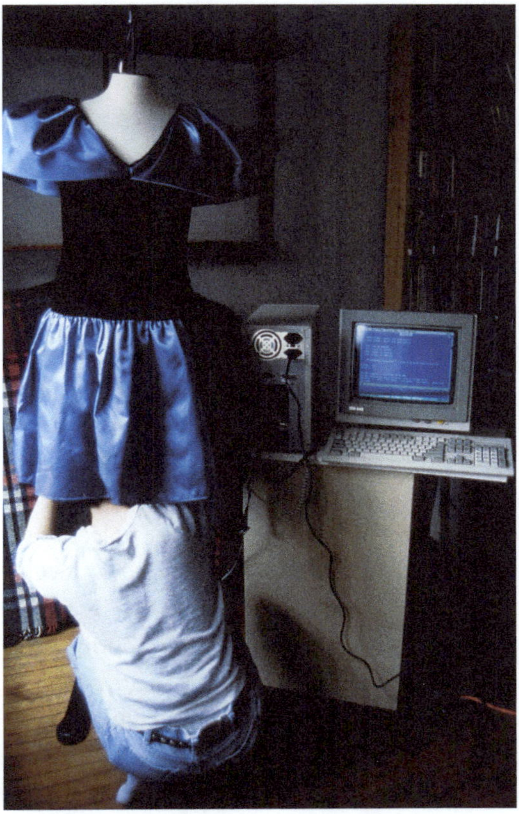

Figure 3.3.2
Nancy Paterson working on the *Stock Market Skirt*. Courtesy Nancy Paterson Estate.

works had primarily been nonnetworked interactive video pieces using videodisc technology.

Rethinking Cyberfeminism

Paterson described her approach as explicitly feminist in her 1992 essay titled "Cyberfeminism."[2] She saw cyberfeminism as a response to the "ecofeminism" of theorists like Camille Paglia who drew a tight link between femininity and nature. In the case of ecofeminism, feminine "instincts" for nurture and a holistic knowledge of nature were essential to overcoming paternalistic social biases. This approach, in Paterson's view, was an untenable retreat from new technologies that were threatening to steamroll women with default patriarchal and military-industrial values. She wrote that "new electronic technologies represent a magic circle from which women have been traditionally excluded."[3] The *Stock Market Skirt* is an important reminder that women and electronics have been historically entwined through industrial labor during World War II, the female-dominated field of telephone operators, and the textile-like craft of wiring early computer systems. The term "computer" was unequivocally feminine, referring not to a thing, but to a woman whose labor was responsible for calculating.[4]

Stock Market Skirt, like the concept of cyberfeminism itself, reclaims electronic technologies for women. "Sex, danger, women, and machines: the plot of virtually every futuristic sci-fi movie in which women play any role at all," Paterson wrote. "Cyberfemmes are everywhere, but cyber feminists are few and far between."[5] Created in a different context, *Stock Market Skirt* might resemble a sexist chindogu device—an objectifying motivator for male business executives to raise the company stock price at the expense of female decency. Instead, *Stock Market Skirt* mocks the straight male fantasy of money being an aphrodisiac to women. In so doing, Paterson intervenes in the narrative of external forces controlling women's clothing and bodies. *Stock Market Skirt* renders this dynamic in a playful, futuristic, and engaging way. As Randy Lee Cutler put it, "*Stock Market Skirt* makes manifest the libidinal economy of technology, money, and gender . . . driven by the collective unconscious of human greed and illuminated through an appetite for fashion."[6] It does not explicitly empower or suppress but draws a link between the collective logic of finance and gendered clothing. In the process, it wrestles control from the dominant narratives around technology as an empowering force.

Stock Market Skirt gracefully builds a metaphor between the accelerating availability of information on the Internet in the 1990s and the hunger of fashion to continually reinvent itself. Paterson's dress hemline slowly falls and rises proportionally to the stock value, directly referring to the theory put forward sociobiologist Desmond Morris and others that fashion and economics are proportionally linked.[7] Morris proposes that in times of crisis and deflation, hemlines lower and colors disappear. In times of growth and at the height of the business cycle, skirt hemlines rise to show more skin, and clothing becomes more vibrantly colored.[8] *Stock Market Skirt* electronically couples stock prices with this "hemline index" and shifts its time scale into an instantaneous interaction in order to present an accelerated view of what Morris and Gaus saw slowly unfold over the course of seasons and decades.

The Gendered Space of Telerobotics

Clothing is, as sociologist Georg Simmel reminds us, part of ongoing conversations about where power lives.[9] Paterson's feminist perspective advanced telerobotic art beyond a "remote control" model by drawing attention to assumptions of masculine dominance. *Stock Market Skirt* is a straightforward counterdemonstration of how new technologies often claim to be "for everyone" but the general user embedded in industrial design is usually a white, upper-class man. It is not a telepresence artwork that transmits human presence—instead, it physicalizes and embodies *data* in real time. Several robotic telepresence projects with human operators over the Web had been deployed before,[10] including Ken Goldberg's *Mercury Project*—the first to enable individual users to remotely view and manipulate the physical environment over the Internet in September 1994.[11] However, Internet telerobots prior to 1998 were projects led by men and were predominantly human-controlled. These Internet telerobots were built as actuators to overcome distance, in which the technology functioned as a personally empowering tool.

Paterson's work, by comparison, amalgamated the handmade with electronic technology and brought together craft and electronics in a unified piece. The software raised and lowered the skirt by way of a stepper motor mounted on the underside of the dressmaker's mannequin.[12] The motor coupled to a round spool that wound up eight rope-like lines threaded through a cascade of hoops sewn on the inside of the dress seams (figure 3.3.3).

As a result of its physical construction, the hemline of *Stock Market Skirt* not only goes up and down; the dress uniformly gathers up, ruffles, and billows as if performing a ruched-style of skirt-lifting curtsy. Randy Lee Cutler suggests that by lampooning the gender dynamics of technology, *Stock Market Skirt* "engages with the implications of new media on women" and explores how "marginalized groups might wrestle control from the dominant techno-narratives."[13]

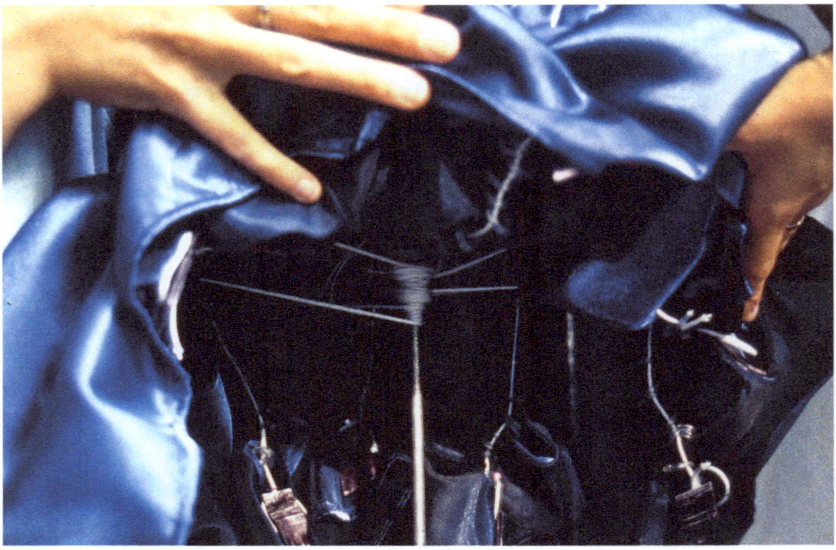

Figure 3.3.3
Detail of the skirt-raising shaft mechanism inside *Stock Market Skirt*. Courtesy the Nancy Paterson Estate.

Interactivity and Data Visualization: The Internet as a Living Thing

Stock Market Skirt is notable because it is the first data-driven Internet-controlled robotic project to use live data off the web. Other projects already discussed in this book, like the *Telegarden* from the introduction, enabled individuals to remotely control a device in a physical location.[14] My version, which I called *Interface*, went against the standard approach of a single user—I encouraged unknown users to make a mess. *Stock Market Skirt* was even more "feral" because it was the first "userless" Internet telerobot, an example of what Eduardo Kac refers to as "telepresence art" but driven by data, not humans.[15]

It is important to note that Paterson's artwork is not interactive in the usual sense of the term—and this is clearly its strength. "Interactive" commonly refers to a human user pushing a button and getting an immediate response. Interactive art tends to refer to artwork that a museum visitor communicates with. Instead of visitors, Paterson's project "interacts" with an aggregate of financial data retrieved in real time from the Internet (figure 3.3.4). In this sense, Paterson's art can be viewed as a critical

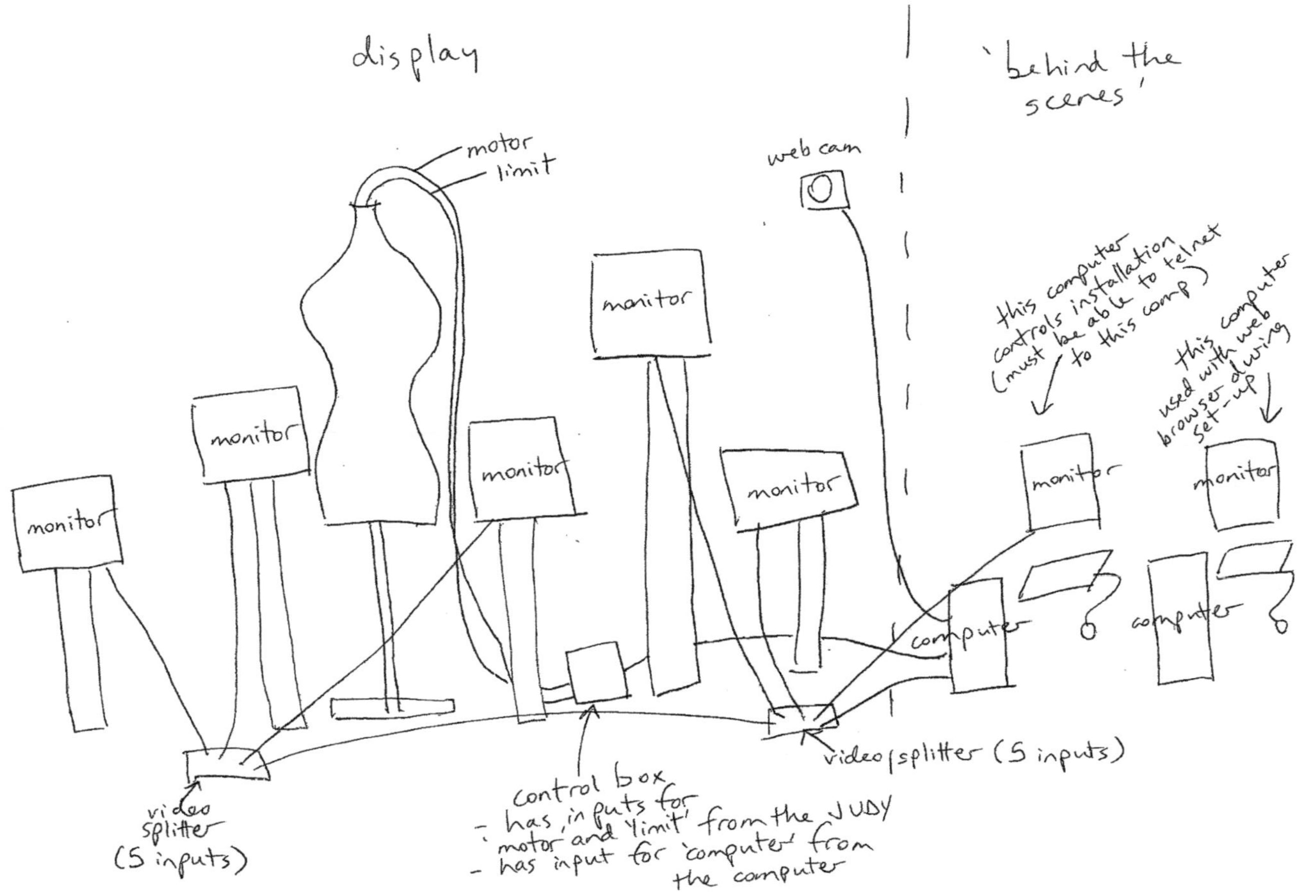

Figure 3.3.4
A hand-drawn schematic sketch from 2017 of the technical components that are part of the *Stock Market Skirt* system by Nancy Paterson. Courtesy the Nancy Paterson Estate.

commentary foreshadowing wariness about what is now called "big data." As danah boyd and Kate Crawford remind us, big data is not universally empowering.[16] Rather, as it is interpreted and creates new knowledge, it threatens to pave over the contextual and cultural meanings of what it means to be human. *Stock Market Skirt* throws these discontinuities of modernity into stark relief: its physical human shape makes forgetting the body impossible. In this way, Paterson foreshadowed a now-common perspective that the Internet is a living, dynamic system with complex emergent behaviors. The Internet is a living thing, and Paterson's work is one of the first projects in history to physically manifest this.

McLuhan's notion that technology is an "extension" of our senses can be taken almost literally in the standard understanding of telerobotics.[17] McLuhan stated that "any medium whatever is an extension, a projection in space or in time, of our various senses"[18]—and similarly that "the wheel is an extension of the foot, the book is an extension of the eye, clothing, an extension of the skin, electric circuitry, an extension of the central nervous system."[19] However, Paterson's cyberfeminist perspective on the gendering of technology gives us a far different vantage point on the idea of extension as a form of control. Paterson's artwork, however, flips the idea of empowering control that permeates McLuhan and the history of early Web-based telerobotic devices like Goldberg's *Telegarden*, *Stock Market Skirt* couples money with sexual desire, accelerating the fashion cycle to ten-second increments. *Stock Market Skirt* reminds us that while new technologies often empower us, they can also control and subjugate us. In other words, elites can use new technology as a means to gain more control, while those without power can often experience new technology as a controlling force. Paterson's project is a metaphor for who is regularly empowered by technology—financial investors and capitalists—and who regularly gets their skirt unexpectedly "lifted up" when the technologies are used against them—women.

In contrast, scholars like Sadie Plant also remind us that the bodies of women are integral to the physical making of technology: "When computers were vast systems of transistors and valves which needed to be coaxed into action, it was women who turned them on. When computers became the miniaturized circuits of silicon chips, it was women who assembled them . . . When computers were virtually real machines, women wrote the software on which they ran. And when computer was a term applied to flesh and blood workers, the bodies which composed them were female."[20] Female labor is clearly integral to the materialization of high technology, and gender remains a strong topic of exploration in a wide variety of DIY practices in history.

Bodies and Wearing Your Politics

Paterson's groundbreaking creation of the *Stock Market Skirt* helps us appreciate the role of the feminist concept of embodiment in DIY electronic art. To understand this work, we must regard clothing as a cultural barometer of society. Clothing is also like a form of interpersonal conversation. Diana Crane, a sociologist at the University of Pennsylvania, believes that clothing is "how people in different areas have perceived their positions in social structures and negotiated status boundaries."[21] In 1995, for example, Anne Balsamo argued that popular media had intertwined technology and the body to a point where "machines assume organic functions and the body is materially redesigned through the use of new technologies and corporeality."[22] In other words, our contemporary era is one where the body is less a fixed part of nature and more a mix of the organic, technological, and cultural.[23] The result of this intermingling is that clothing and technology can be powerful statements of feminist art. Paterson's project also engages with the deeper implications of new media, as it was one of the first devices connected to the Internet that went beyond the cliché of the web being merely an empowering place.

Instead of portraying telepresence as an empowering extension of the body, Paterson's work shows us an Internet that is repressive and controlling. Her insistence on having a body—represented as a physical dress on a dressmaker's mannequin—although contemporary with other cyberfeminists and technofeminists, ran counter to prevalent discourse around the Internet that promoted the benefits of disembodied online interactions. Writers like Howard Rheingold saw Internet and VR technology as a prosthesis, enhancement, and replacement for biology and all of its problems, which helped feed a frothing tech industry eager to sell products.[24]

Other technology scholars, like Sherry Turkle for example, saw significant potential in the anonymity of online interactions. Turkle elaborates: "You can be whoever you want to be. You can completely redefine yourself if you want. You don't have to worry about the slots other people put you in as much. They don't look at your body and make assumptions. They don't hear your accent and make assumptions. All they see are your words."[25] In this vein, on July 5, 1993, the *New Yorker* ran what would become its most reproduced cartoon ever: a drawing of a dog using a computer and telling another dog, "On the Internet, nobody knows you're a dog."[26]

The problem with transcending the body and leaving it behind, according to Anne Balsamo, is that physical bodies continue to matter. The dream of leaving the body behind neglects how most of us are tied to the corporality of our bodies: Black bodies, women's bodies, racialized bodies, and transgendered bodies are profoundly intertwined with identity. The desire to escape the body, in this sense, can be seen as a deletion of the embodied experience that is so important to many. In other words, "Masculinist dreams of body transcendence . . . signal a desire to return to the 'neutrality' of the body, to be rid of the culturally marked body."[27]

Balsamo continues by explaining that a cyborg-like blurring of the boundaries between the "biological" and the "technological" does not erase gender.[28] True, it is no longer a fixed, stable, or "natural" thing, but physical bodies clearly do still exist since the invention of the Internet, and bodies typically are gendered. Instead of simply deleting gender and replacing it with genderlessness, the technological upheaval of the Internet in the 1990s blurred its lines and helped expand it into a gradient.

Donna Haraway's *Cyborg Manifesto* is a popular touchstone for technofeminist thought on embodiment that charts a route between the usual utopian and dystopian poles. In it, Haraway proposes that humans and technology have become embodied and hybridized. The twentieth century saw boundaries get blurred—including between humans and animals, the animal-human and the machine, and the physical and the nonphysical.[29] In response, Haraway proposes, evolution and technology demand a nonbinary perspective: "Cyborgs and companion species each bring together the human and nonhuman, the organic and technological, carbon and silicon, freedom and structure, history and myth, the rich and the poor, the state and the subject, diversity and depletion, modernity and postmodernity, and nature and culture in unexpected ways."[30] Haraway goes on to argue that the body cannot simply be transcended. Instead, a body wholly instantiates or "genders" technology.[31]

As Hélène Day-Fraser and Keith Doyle point out, clothing is often less of a unidirectional fashion "statement" and more of a multidirectional fashion "conversation."[32] This conversation is intricately linked to representations of gender. In academic terms, this conversation can operate as "participatory research methods"—with Day-Fraser and collaborators highlighting how clothing can function as a cultural probe (Bill Gaver), a cocreation (Elizabeth Sanders and Pieter Jan Stappers), as critical design (Tony Dunne and Fiona Raby) and as a form of heuristic and mindful inquiry (Dave Hiles; Valerie Bentz and Jeremy Shapiro).[33] Unfamiliar clothing often results in visual and mental engagement, and this can be a point of discussion or learning for the wearer or viewer. Paterson's custom-built clothing operates in this way and explores the multifaceted gradient of gender in the process.

Wearing politics can be seen far beyond Paterson's work, however. This style of work is reminiscent of the fashion technology Kat Jungnickel highlights in her research on Victorian bicycle wear for women, namely the Bygrave Convertible Skirt.[34] Similar to the *Stock Market Skirt*, the Bygrave Convertible Skirt has a pulley system built into its seams that allows the wearer to hike it up in order to facilitate bicycling, which in the 1890s was celebrated as a vehicle of women's liberation (figure 3.3.5). As with Paterson's invention, Bygrave's skirt mechanically rises and falls as the environment changes—and both can be seen as industrious and innovative inventions that place the topic of gender in design on center stage.

Conclusions on *Stock Market Skirt*

Paterson's core concept behind her initial conceptualization of cyberfeminism is still valid: "Whether directly or indirectly, issues of economics, class, race, nationality, personality, and gender are driven and defined by new electronic technologies." And her cyberfeminist stance is particularly important for the purpose of this book, which explicitly pushes back on the patriarchal overtones of maker culture. Paterson believed cyberfeminism would evolve into networked feminism, but died to soon in 2018.[35] She left us to imagine how she might have created art in an era when the #MeToo movement has profoundly influenced how feminists act collectively to disrupt systems that are always layered and tangled in patriarchy.

Bringing this back into a larger discussion about DIY production, *Stock Market Skirt* is particularly good at reminding us about the personal

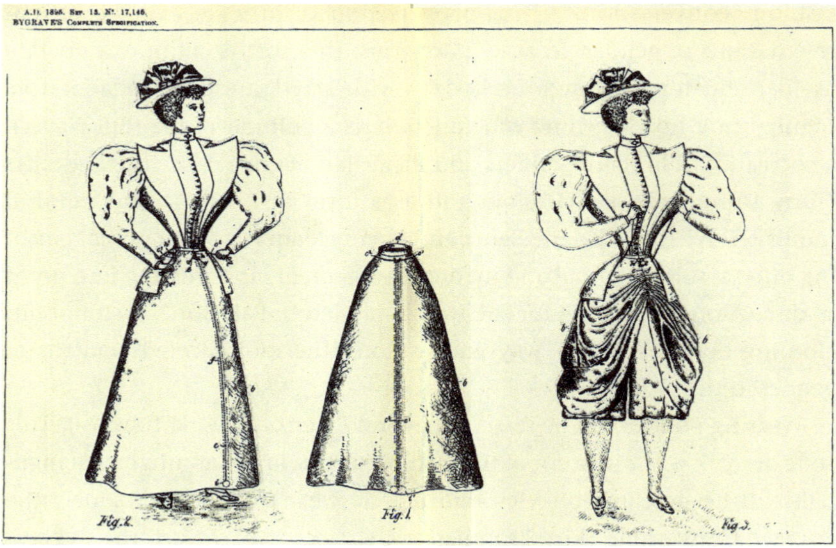

Figure 3.3.5

Alice Bygrave's patent drawing of the convertible pulley cycling skirt from 1895, which splits apart and rises to facilitate riding a bicycle. This patent application is titled "Improvements in Ladies' Cycling Skirts." Canadian, German and American patents were also filed for this invention during December 1895. *Source*: British Patent 17145, submitted November 1st 1895, accepted December 6th 1895.

components of DIY work—that projects often deeply reflect personal attributes. Gender is an overwhelmingly significant factor in life, and therefore is a significant factor in the things people do. Although the term "cyberfeminism" may sound dated, Paterson's core concept still holds water today: "At the heart of Cyberfeminism is the notion not to accept as inevitable the current applications of new technologies which impose and maintain specific cultural, political and sexual stereotypes, and [the belief] that the empowerment of women in the field of new electronic media can only result from the demystification of technology."[36] Her projects explore how marginalized groups can seize control from the dominant narratives around technology, despite the core biases built into systems and technologies.

Considering again the research of Kat Jungnickel, the Bygrave skirt highlights how women shaped clothing to expand their urban lives; in contrast, Paterson's skirt is more ambiguous and less empowering. Victorian-era female inventors used customized skirts to explore the city and to control the relatively new technology of the bicycle, but Paterson's skirt is more like the inverse impulse, with new technology autonomously controlling one's clothes. Jungnickel's work also reminds us that patriarchy is the predominant paradigm of capitalism, and that the concerns of women are often a do-it-yourself initiative that happens outside the core institutions of industry. Both projects remind us that do-it-yourselfness is not a panacea, but a core methodology in women's struggle for power.

In summary, personal DIY, as reflected in Paterson's work, shows strong feminist influences, particularly the notion that the personal is political. Electronic art projects like Paterson's *Stock Market Skirt* bring the intermixing of the topics of gender, technology, control, telerobotics, communication, and bodies into view. The independently produced project adds a useful counterpoint to the aspirations of those developing commercial telerobotic devices: the Internet and technologies in general do not simply free us from limitations. DIY often reminds us of the limitations of humans and technologies, and *Stock Market Skirt* reminds us that technology often imposes one-way control, which is often gendered. Paterson's project is particularly useful for taking the industrial technology of Internet telerobotics and opening it up to discussions around gender. As Anne Balsamo puts it, design has always been a collective process where the gendering of technologies can be reproduced or disputed.[37] In this way, electronic and DIY practices reflect the experiences through the gradated lens of gender, cast by artists as sites of inventive technocultural production to contest dominant narratives in society.

Theme 4 Anti-institutional Disobedience

Disobedience and DIY Electronics: A Thematic Overview

DIY production, which normally occurs outside of a formal institution, is regularly used to launch criticism against established organizations including governments, associations, and companies. The nimbleness of individuals doing their own thing often results in people designing and building things in order to rail against or protest larger organizations in control.

Struggle for a voice or for political power often results in creative forms of making objects. While financial and resource limitations are an overarching variable in the DIY mindset, not all work in this field is a jugaad hack. Although makeshift materials and resources are prevalent, the core thrust of this work is institutional rebellion—work that puts its "body against the barricades" in protest, so to speak.

In society, political protest is a fascinating site of creative communication and making. In addition to protest signs, individuals and organizations regularly launch into more creative forms of messaging and campaigning. For example, in 2014 the Victoria and Albert Museum (V&A) in London opened a major exhibition that highlighted how everyday objects can be inventively repurposed for urban political protest.[1] The show, titled *Disobedient Objects*, showcased a collection of activist ephemera from the 1970s to the present and gave an indication about the breadth of this field. The exhibition also explored the do-it-yourself methodologies of protesters, showcasing instructions for tear gas masks made out of plastic drink bottles, pamphlet bombs, homemade shields against riot police, and covert stencils for spray painting political graffiti.[2] The V&A recapped the exhibition as follows: "On display were arts of rebellion from around the world that illuminate the role of making in grassroots movements for social change: finely woven banners; defaced currency; changing designs for barricades and blockades; political video games; an inflatable general assembly to facilitate consensus decision-making; experimental activist-bicycles; and textiles bearing witness to political murders."[3] Although the artifacts spanned the globe—from Turkey, Greece, Venezuela, Australia, Spain, Syria, Britain, the US, the UK, and beyond—all highlighted the inventive ways that individuals regularly use and modify conventional objects into unconventional tools to challenge institutions.

Do-it-yourself approaches to electronics can also work in a similar spirit to the works in *Disobedient Objects* at the V&A: technology is often deployed in the process of social struggle. Antagonism bubbles up in creative ways. Like plastic soda bottles cut apart into gas masks, technology of different kinds can be cut apart and readapted into tactical political applications. This section of chapters continues with the V&A *Disobedient Objects* initiative, which fleshed out an "under-examined area of the art and design of object-making within social movements, a people's history of art and design from below."[4]

Design for disobedience is not a new concept, however. At least a dozen related labels that describe protest-oriented applications of media art and experimental design exist. Terms include *critical engineering* by Julian Oliver, Gordan Savičić and Danja Vasiliev, *critical making* by Matt Ratto and myself, *critical technical practice* by Phil Agre, *reflective design* by Phoebe Sengers, *critical design* by Anthony Dunne and Fiona Raby, *adversarial design* by Carl DiSalvo, *interrogative design* by Krzysztof Wodiczko, and many others.[5] Each concept has its own audience, attitude, and method, but the general purpose is the same: to use art and design methods with technology to creatively highlight how structural problems exist in society.

Out of this array of choices, the closest relatives to the disobedient styles of DIY practice discussed in the pages ahead are Interrogative Design as articulated by Krzysztof Wodiczko[6] and tactical media as described by Geert Lovink, David Garcia, and others.[7] Wodiczko's concept of Interrogative Design is particularly worth mentioning here because of its powerful but seldom-mentioned impact on this "disobedient" field of work, especially on those coming from the world of contemporary art practice and design disciplines. In comparison to critical design, interrogative design by Wodiczko has the transformative twist of actually deploying intentionally provocative devices in the real world.[8]

Disobedient DIY: Wodiczko's Interrogative Design

Krzysztof Wodiczko is a Polish artist best known for his architectural-scale slide and video projections on prominent building facades and monuments. These projections typically engage with topics around trauma, war, conflict, memory, and communication—and they work to promote public dialogue around marginalized citizens and cultural issues that are rarely given attention by the commercial design industry.[9] He frequently explores topics of concern to the homeless, immigrant, alienated youth, and war veteran communities.[10]

In 1994, while a professor of visual arts at Massachusetts Institute of Technology (MIT), he released a short, two-page essay titled "Interrogative Design" that summarized his approach to art. Wodiczko's essay is notable for several reasons. First, it tramples core distinctions between art and design, jumping straight to the idea that artists *are* designers. By comparison, Dunne & Raby's Critical Design has more of a disciplinary barrier between art and design. Although Wodiczko works as an artist, he clearly frames his work as engineering design and invites other artists to take on the task of industrial design. The disciplinary interconnections are important and worth noting: Wodiczko's disciplinary provocation is to combine art along with hands-on technology in order to do industrial design in partnership with cultural issues.

The way this is done is by asking questions—"interrogating" in other words. Wodiczko's document opens with a question-focused definition of "*interrogative*: 1. Of, pertaining to, or of the nature of questioning; having the form or force of a question. 2. Of a word or form employed in asking questions." A key here is that Wodiczko discards the traditional frameworks of art being representative, visual, or even aesthetic.

Instead, Wodiczko leaps into the task at hand that studio work needs to work against the "painkillers of optimistic design fantasies." This is done by combining technology design with the participation of social art practice in public space to raise well-crafted questions about why and how different social difficulties exist. The well-crafted questions are key and are aimed at agitating the public into innovative reevaluations of complex social problems. "Interrogative design questions the very worlds of needs of which it is born. It responds interrogatively to the needs that should not, but unfortunately do, exist in the present 'civilized' world. In the unacceptable world, interrogative design should present itself and be perceived as unacceptable."[11]

An example of Wodiczko's approach is seen in his *Homeless Vehicle* project, a series of vehicles he started designing and producing in the late 1980s for individuals experiencing homelessness (figure 4.1). One of his custom-fabricated designs resembled a shopping cart that expanded out into a sleeping capsule, and it also featured a storage area for personal belongings, a washbasin that doubled as support for a table, and a bin to hold cans and bottles for recycling (figure 4.2). Although the project does not incorporate DIY electronics, it is worth considering here to gain a better understanding of Wodiczko's influential approach to disobedience.

The *Homeless Vehicle* is disobedient design because it *intentionally* proposes an *imperfect* solution to a problem. It is a surface-level fix for homeless individuals but is a bit akin to the Yes Men's *Survivaball* suit, an odd,

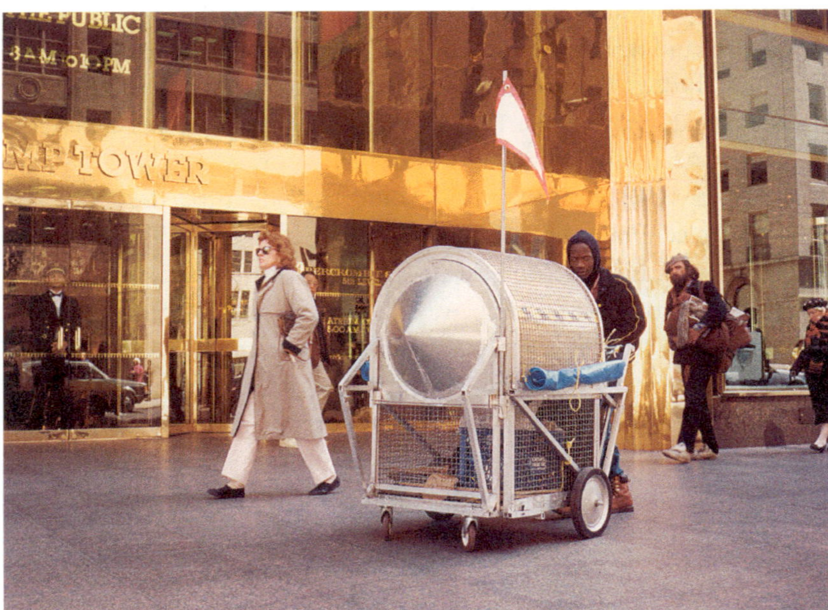

Figure 4.1
Homeless Vehicle in New York City in front of Trump Tower by Krzysztof Wodiczko, 1988–1989. Photo Hirshhorn © Krzysztof Wodiczk. Courtesy of Galerie Lelong & Co., New York.

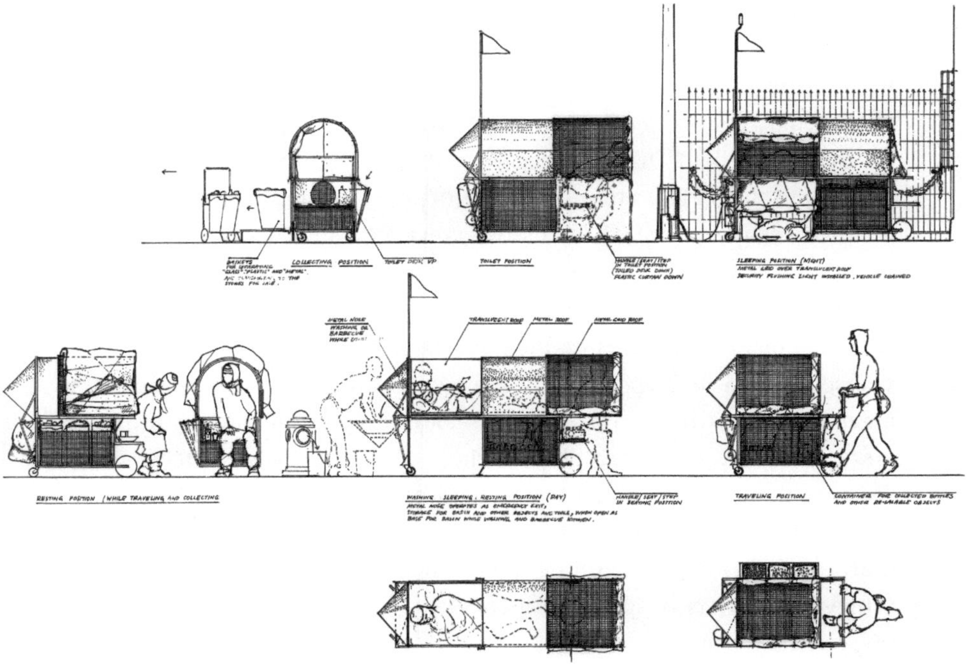

Figure 4.2

A hand-drawn diagram of the *Homeless Vehicle*, showing its various functions including as a recycling system, toilet, and bed. Courtesy Krzysztof Wodiczko.

inflatable suit that looks like an absurdly over-inflated tardigrade and promises to protect the wearer from climate change.[12] Wodiczko's vehicle is similarly an intentionally ludicrous piece of engineering, clearly wallowing in the ambiguity of whether it is a good idea or not. In other words, the whole thrust of the project is to engineer a creative solution that twists the public into actually talking about the issue of housing the homeless. It is intentionally *both* a partially absurd and a partially functional engineering solution, and this uneasy frisson creates meaningful public discussion around the complex social problem. It operates similarly to chindogu in terms of being an intentionally awkward tool or solution. This unsettling component of half disobeying the engineering mindset pulls the public into the debate; it challenges individuals to participate in finding a better solution.

Disobedience and Tactical Media

In addition to the concept of Interrogative Design, the concept of "tactical media" is also clearly worth exploring as a component of adversarial and disobedient DIY practices. As a term in the media arts, "tactical" bubbled to the surface around 1993 through a coalition of cultural groups in Amsterdam who called themselves Next 5 Minutes (N5M). They organized an event titled *"Tactical Television,"*[13] echoing and extending earlier activist uses of the television medium by media artists,[14] like *"Guerrilla Television."*[15] Tactical media is defined as "what happens when cheap 'do it yourself' media made possible by the revolution in consumer electronics, are exploited by those who are outside of the typical hierarchies of power and knowledge."[16] Similarly emphasizing N5M's move between tactical *television* to the broader tactical *media*, Critical Art Ensemble (CAE) describes tactical media as "a critical usage and theorization of media practices that draw on all forms of old and new, both lucid and sophisticated media, for achieving a variety of specific noncommercial goals and pushing all kinds of potentially subversive political issues."[17] This work de-emphasizes the traditional role of the artist within a gallery system, instead envisioning a hybrid cultural producer that melds the roles of artists, scientists, technicians, craftspersons, theorists, and activists.[18] From the start, tactical media took amateur practice seriously. Amateurs leverage being outside of policy, funding, and institutions. As a result, they are free to intervene, smash, and subvert dominant paradigms.

Critical Art Ensemble, Next 5 Minutes, and many others borrowed the term "tactical" from Michel de Certeau's *The Practice of Everyday Life*, which examines how individuals modify and personalize cultural objects, practices, and spaces.[19] In this influential work, de Certeau makes a distinction between *strategies* and *tactics*. *Strategies* are how formal institutions organize and exert control, while *tactics* are the methods that individual users navigate around and use to subvert official objects, spaces, and practices. Activist, punk, and creative components of the community converged together around the smash-and-grab DIY attitude of "tactics" and "tactical media" in the 1990s. In 1997, Geert Lovink and David Garcia published the essay "The ABC of Tactical Media," and as a result, the term gained significant traction in the media arts and activist circles.[20]

Rita Raley defines tactical media as a kind of collective performance. In tactical media, there are not clearly delineated classifications like "builders," "curators," and "audiences."[21] Instead, the primary focus is on the general category of "participants." The goal is not to build an electronic product or object, rather, the goal is for participants to destabilize unfair organizations, powers, and systems of discrimination. In this way it shares common ground with Wodiczko's Interrogative Design as being more dialogue focused than engineering-oriented problem solving. Tactical media does not propose a grand or systematic revolution or provide any explicit structural transformations. It grabs ideas and symbols to hack and exploit for institutional criticism.[22]

Disobedient DIY projects discussed in the next section of chapters use electronics as a form of protest to intervene in the political, economic, and cultural aspects of technology in culture. According to Rita Raley, a key component of this tactical work is the intervention and disruption of a "dominant semiotic regime."[23] In other words, the work often scrambles and subverts images and messages in order to create a temporary space for reflection and critical thinking. It regularly uses symbolism to attack the neoliberal condition, which she defines as an ideology that stipulates that "financial markets should operate unfettered and that state intervention or regulation of any kind is anathema."[24] Tactical projects are also often "exploits" that produce unexpected moments of political potential. Galloway and Thacker explain that exploits are not centered on "changing existent technologies but instead involve discovering holes in existent technologies and projecting potential change through these holes."[25] Disobedient DIY uses improvisation to respond to immediate circumstances and lacks a larger strategy. That is, it includes "projects that people do opportunistically—seizing temporarily available or unclaimed resources."[26]

This section of chapters strives to carve out different ways that DIY work flexes its disobedience. It opens with chapter 4.1, titled "Disobedience and *Robot K-456*: Wabi-sabi, Electronic Arte Povera, and Beautiful Mistakes." Next, chapter 4.2 is "Disobedience and *Hairbrain 2000*: Burlesque Technologies, Highlowness, and Neoretroism." The thematic section on disobedience ends with chapter 4.3, titled "Disobedience and *Feral Robotic Dogs*: Hardware Activism, Communities, and Planned Obsolescence."

In the process, these chapters explore rebellion against formal institutions of art, against the idea of technological progress, and against the technology manufacturing sector. Thomas Snow "reminds us both that design and political art needn't orientate itself completely to the market, and that a material, pluralistic, and diverse democracy is a commodity worth fighting for."[27] Through these examples, we will explore how creative DIY builders adapt, use, create and fight with technology in spectacularly disobedient ways.

4.1 Disobedience and *Robot K-456*: Wabi-sabi, Electronic Arte Povera, and Beautiful Mistakes

Artists and designers often embrace a DIY methodology or aesthetic to challenge an established system. Like the antiestablishment attitude woven into the punk movement in 1970s London, DIY often has elements of an antiestablishment "fuck you" agenda aimed at criticizing refined culture or commercial enterprises. In punk subculture, DIY stood for a grassroots democratization of making music and clothing, but antiestablishment DIY movements have been prevalent long before the 1970s.[1] Johnny Rotten of the Sex Pistols acknowledges a long line of antiestablishment predecessors to punk sensibilities, for example, including the "'teddy boy' movement, the skinheads, the mods, and the rockers all opposed to the hold that the high street had on fashion."[2] Contemporary art also taps a continual flow of antiestablishment sentiment that operates generationally. Younger generations often take older, established artists and their institutions as their targets. Intentional use of nontraditional materials and methods provide useful ammunition in attacking the older art establishment. In this way, electronics clearly plays an adversarial role in the field of contemporary art. As a medium, it routinely challenges the mass understanding of art as merely paintings and drawings easily presented by curators and gallerists.

The homebrewed aesthetics of technical DIY projects—whether in music, art, or electronics—typically clash with the glossy mainstream and often grind against formal institutions. This counterinstitutional attitude can take many forms, including many of the other themes proposed in this book, like jugaad, burlesque, parody, protest, or sacrilege. A more subtle dynamic of critique, however, is a magnification or celebration of the imperfect and handmade quality of the work—the "beautiful mistakes" of a maker's craft. Although several terms can be used to describe this handmade aesthetic—like "rustic" or "folksy"—the Japanese concept of wabi-sabi is incredibly useful to understand and gain an alternate perspective on this component of DIY aesthetics. As an aspect of traditional Japanese aesthetics, wabi-sabi provides a useful building block to rethink how DIY objects and technologies operate and what social value they hold. Understanding wabi-sabi helps us unlock and utilize an important aspect of the DIY mindset.

Nam June Paik: Origins and Early Work

Korean-born artist Nam June Paik (1932) provided a number of strong examples of antiestablishment DIY electronics in his early career. Paik is best known for his extensive body of video-based installation using tube television sets as sculptural objects, including works like *TV Buddha* (1974/1982) and the *Family of Robot* series (1985). In 2000, John Hanhardt wrote for the Guggenheim that "no artist has had a greater influence in imagining and realizing the artistic potential of video and television."[3] For the purposes of our discussion, however, Paik's earlier work using more rudimentary electronic circuits in the 1960s is especially relevant as an example of how the medium of electronics can be deployed as a tactical tool to critique institutions.[4] As art historian Edward Shanken notes, Paik worked in relative DIY-style obscurity until the 1980s.[5]

Paik's early work had many influences, but John Cage's experimental work using everyday objects, chance and artistic performativity within the framework of music is notable.[6] As a "gadfly of the musical avant-garde," Paik also aligned with George Maciunas and many early Fluxus artists in their efforts to decommodify and de-emphasize finished art objects, instead focusing on the creative and performative artistic process.[7] In particular, Paik was intent on waging war against the rarefied art object, particularly those made by the older generation of pop artists and abstract expressionists like Jackson Pollock.[8] An example of the intermingling of Cage's influence and an interest in unrarefying art objects can be seen in Paik's performance at one of the first Fluxus events titled *Neo-Dada in der Musik* in Düsseldorf Germany on June 16, 1962. At the event, Paik performed his piece *One for Violin Solo*, in which he slowly raised a violin over his head over the course of five minutes, then smashed it down on a table in front of him with a single cacophonous blow.[9]

Paik had been experimenting with electronics before performing *One for Violin Solo*, however. His work had incorporated recorded sound since at least 1959, when in the project *Hommage à John Cage* he mixed performance with a spliced montage of "piano playing, screaming, bits of classical music, and sound effects."[10]

Robot K-456: Metal Bars, Electronics, Spinning Foam Rubber Breasts, and a Floppy Glove

Paik had also been experimenting with electronic control systems originally designed for wireless radio-controlled airplanes since at least 1960.[11] This interest in radio-controlled systems expanded into starting the development of *Robot K-456* in 1963 and continued through 1964. Structurally, *Robot K-456* is 183 cm (6 ft.) tall and has a frame constructed out of metal bars that are haphazardly bolted together to form a rough silhouette of a skinless human. It has a box head, lopsided, sticklike arms, an asymmetrical torso, and heavy, blocky legs.[12] As an assemblage, the system resembles a golem that spontaneously formed out of a junk pile of steel bar, wires, and scrap electronics (figure 4.1.1).

The technical details of the machine are worth noting due to its peculiarity: its rectangular head has a large, black, speaker coil for a mouth, capped with a bent, tin pie-plate-looking hat. The two arms, uneven in length, hang from a horizontal bar that acts as shoulders. A motor is perched on top of the left shoulder, enabling the left arm—and the floppy glove dangling from the end of it—to lift and wave. The torso has a pair of spinning foam rubber breasts that are placed asymmetrically on the body.[13] Underneath, near the center of the abdomen, is an electric fan. Spindly wires are strung across and around the entire structure. Asymmetrical legs constructed out of L-channel metal bars sit on top of two skateboard-like feet, with the bulkier left leg containing the majority of the electronic components in the system, in addition to batteries and other components that appear to constitute the robot's electronic "brain" or control center.

Robot K-456, reinforcing the collaborative nature of many DIY projects, was not built solo. The robot had technical help from both Paik's brother Ken and the engineer Shuya Abe. While, in general, some DIY projects are stubbornly single-authored creations, most have significant input from others. Shuya Abe, for example, initially started working with Paik on *K-456* and their collaboration continued for much of Paik's lengthy career. As I

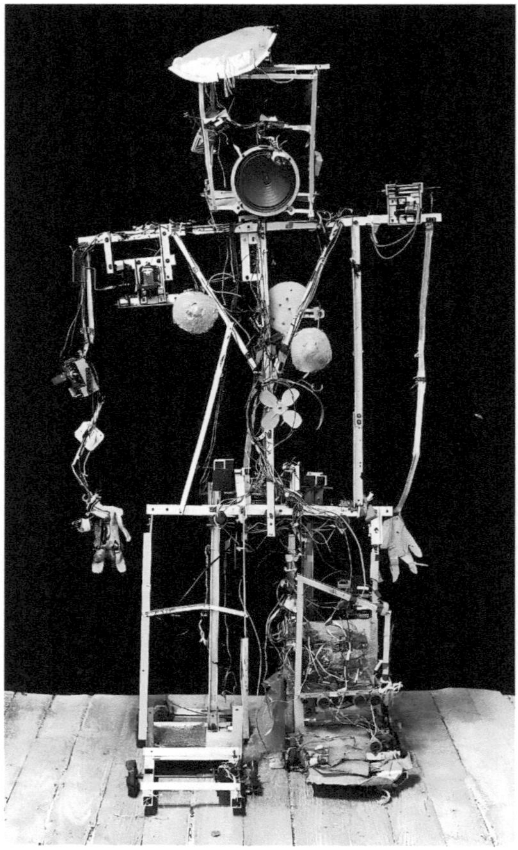

Figure 4.1.1
Robot K-456 (1965) by Nam June Paik. Photo Peter Moore, © Northwestern University. Courtesy the Charles Deering McCormick Library of Special Collections, Northwestern University Libraries.

previously described in the "A Definition of DIY: Do+It+Yourself" chapter, DIY should not be confused with artistic isolation. Rather, DIY practice is more an attitude of stubbornly following one's own direction and operating outside of established institutions and capitalistic demands.

Paik's robot was named after a relatively unknown Mozart piano concerto from 1784, Köchel catalog number 456, abbreviated "K-456."[14] The name jabs at the establishment of classical music and intentionally taunts us. It provokes observers to equate this junky and haphazard construction with "not just music but classical music that should be taken no less seriously than a work by Mozart"—metaphorically similar to the sacrilege of *One for Violin Solo* smashed as a classical music performance.[15]

K-456 Operation: "I wanted it to kick you and then move on"

The robot premiered on August 30, 1964 in New York at the Second Annual New York Avant-Garde Festival in Judson Hall and outside through the surrounding streets (figure 4.1.2).[16] The head-mounted speaker was connected to an amplified tape player configured to play recorded political speeches.

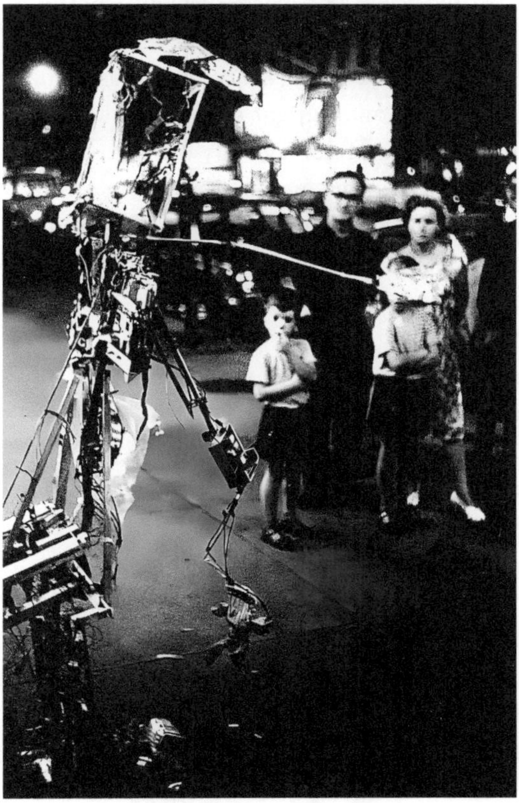

Figure 4.1.2
Robot K-456 performs on a public street in New York City in 1964. Twenty-channel radio-controlled robot, aluminum profiles, wire, wood, electrical divide, foam material, and control-turn out. 72 x 40 x 28 in. (183 x 103 x 72 cm). Photo Peter Moore, © Northwestern University. Courtesy the Charles Deering McCormick Library of Special Collections, Northwestern University Libraries.

This Fluxus-style audio montage included Kennedy's 1961 inaugural address, "Ask not what your country can do for you, ask what you can do for your country" and monologues from Hitler and Winston Churchill—which emphasized the dynamic of how political speeches have their unique form of agency and carve out a life of their own.[17] The robot structure could bow, shuffle along, lift each arm, and wiggle its flat torso. While walking, the robot's feet would regularly roll and bump into each other and need to be manually repositioned. The entire system took four people to operate and at least three separate handheld radio modules with antennae, knobs, and switches to control.[18] The remote-controlled machine walked, talked, and "shit" dried white beans out of a mechanical dispenser from its backside as excrement.[19]

Echoing the emphasis on temporary events and "happenings" of his Fluxus contemporaries, Paik thought of *K-456* as a performative intervention amongst the public and not a gallery-oriented piece of art. "I thought it should meet people in the street and give one second of surprise," he said. "Like a quick shower. I wanted it to kick you and then go on. It was a street-music piece."[20] In other words, Paik imagined his robot as a visually jarring mechanical busker.

Seeking Imperfection: Critiquing Art and Technology via Wabi-sabi and Electronic Arte Povera

Paik's *K-456* inverted the prevailing progressive futurism around humanoid robots in popular culture of the late 1950s and early 1960s; *Robot K-456* was clearly not a glossy, futuristic robot. For example, the popular Robby the Robot in Metro-Goldwyn-Mayer's (MGM's) featured in several science fiction films during the 1950s serves as a clear counterpoint to *K-456*. Robby was a publicly recognized film star crafted in the image of a technologically advanced android: it was a Hollywood portrait of the future, including chrome detailing, blue neon lights that pulsed when the robot spoke, and metallic, vacuum-formed plastic skin (figure 4.1.3). Robby was portrayed on screen as a helpful mechanical servant with dry wit and its own artificial intelligence, perhaps like the contemporary personalities of Siri, Google Home, or Amazon Alexa in humanoid form. *Robot K-456* perverts the smooth, technologically advanced approach of Robby. Robby's progressive futurism is manifested throughout the history of biomimetic robotics,

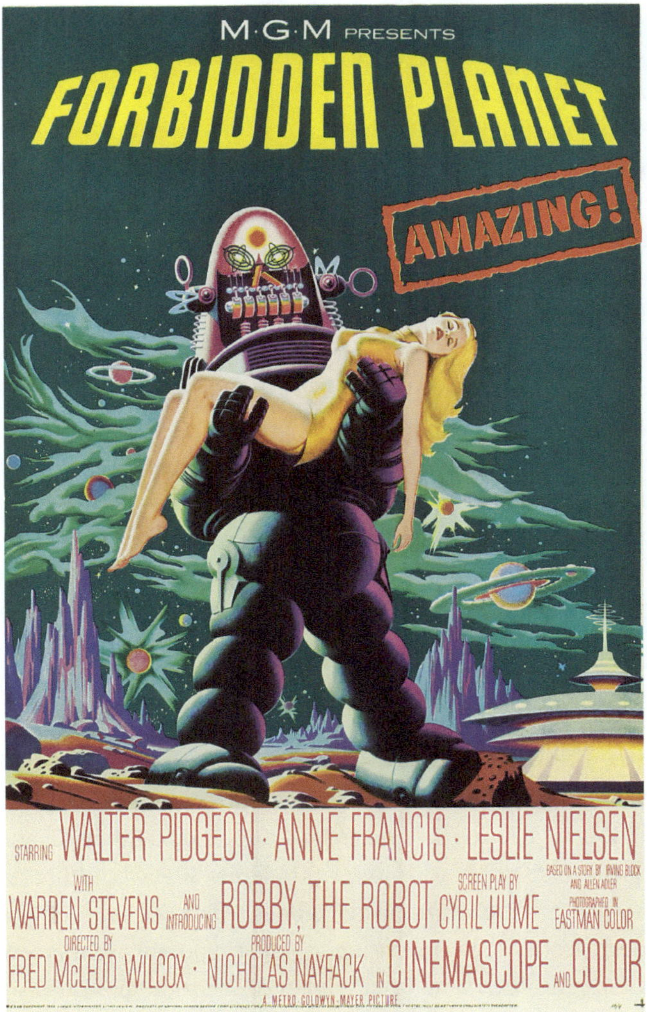

Figure 4.1.3

Robby the Robot is represented as an intelligent and functional mechanical servant in MGM's 1956 film, *Forbidden Planet*. Public domain, courtesy the National Screen Service Corporation.

spanning the intricate eighteenth-century automatons of Jacques de Vaucanson, Fritz Lang's female "menschmaschine" in 1928's *Metropolis*, and characters like Star Wars' C-3PO from 1977.[21]

This "imperfect" approach was a conscious decision of Paik's. When Paik's brother offered to help make a more electronically and mechanically advanced robot that could reliably walk instead of limping, Nam June clearly refused. Melissa Chiu, director of the Smithsonian's Hirshhorn Museum and one of Paik's core contemporary proponents, agrees that Paik clearly did not want a perfect machine.[22] *K-456* intentionally clings to the electronic-scrap-heap aesthetic of its raw materials as a fragile structure, and intentionally embraces inexpensive everyday items like a pie tin and a bra.[23] *K-456* also completely lacks an exterior skin or veneer: it exists as an exposed metal skeleton with all technical entrails exposed. This was clearly a tactical maneuver to embrace "poor" materials as a form of critique—and at the time, using these kinds of materials was still a relatively novel approach within contemporary art.[24] Paik's *K-456* was a critique of at least two institutions: the institution of contemporary art and the institution of advanced technologies. While critiques of the institution of art can be thought of as being in line with the Fluxus sentiment following Duchamp, Dada, or Cage, the institution of advanced technologies can be thought of through the metaphor of Robby the Robot. This certainly extends beyond the Hollywood caricature of robotics, however: the progressive futurism that spawned Robby is also echoed through a cultural belief that advanced technologies are the solution to human problems. Silvia Lindtner and coauthors term this "technosolutionism": a belief that societal problems can be fixed through technological solutions.[25] Most reputable scholars of science and technology studies agree that advanced technologies offer significant gains in certain spheres, but these typically come at a cost—and high tech often shuffles the variables around with initially unforeseen negative consequences predictably arising over time.[26] Yes—advanced technology often improves life—but it does not fix social problems on its own.

As Wulf Herzogenrath notes, in 1965 Paik referred to *Robot K-456* as "poor" art, in opposition to the "rich" Pop Art of the same era, which "seems to have been one of the first references to the concept of 'poor art,' for in 1965 Italian arte povera did not yet exist."[27] Although at the time some critics simply viewed the work in terms of its technological unreliability and lack of looking "state of the art," others saw that Paik's choice of materials and level of finish in *K-456* were intentional: he had "at his disposal perfect modern technology, yet to some extent he [went] out of his way to get by without it."[28] Note that this "poorness" is distinctly different than the concept of jugaad (discussed earlier in reference to Norm White and Doug Back's *Telephonic Arm-Wrestling* in chapter 1.2)—this is more a chosen poverty like Thwaites's toaster as opposed to a lack of resources.[29]

Wabi-sabi is useful in understanding the dynamics of the antisolutionism of *K-456*. The concept originates in traditional Japanese aesthetics referring to a beauty that is "imperfect, impermanent, and incomplete."[30] Although a direct translation into English is difficult, when split apart into its separate words, "wabi" refers to a rustic simplicity that can exist in natural or human-made objects as an understated gracefulness or the beautiful irregularity of a handmade artifact. "Sabi" is more like the patina of time that slowly accumulates through wear, repair, or aging. Together, wabi-sabi can be thought of as a manifestation of Zen Buddhism, highlighting an appreciation of the "flawed beauty" of imperfect things and coming to terms with the evanescence of time.[31] Daisetz T. Suzuki, a Zen Buddhist scholar, describes wabi-sabi as "an active aesthetical appreciation of poverty"[32]—but not in a needy sense of poverty, but more as a simplified doing-with-less and an acceptance of imperfection. This is humility by choice, and not yearning to be something else. Wabi-sabi can be thought of as the beauty of a handmade piece of imperfect pottery, or the elegance of an abandoned and partially collapsed barn on the Saskatchewan prairie.

To circle back to comparing jugaad to wabi-sabi, the former typically comes out of a forced lack of resources and the latter is more of an aesthetic choice. Wabi-sabi's origins are distinctly upper class, which according to Tom Jennings "began, centuries ago, as a reaction of Japanese monied classes' desire for more enhanced, exclusive, and rarefied aesthetic pleasures. . . . Actual poverty is nowhere to be found within wabi-sabi."[33] While both concepts relate to frugality, jugaad is externally imposed constraint and wabi-sabi is internally imposed restraint.

K-456 has many aspects that are clearly outside of the wabi-sabi aesthetic. With its mess of wires and components, the machine is far from minimal. It also does not show a traditional material "sabi" patina. However, the project uses materials honestly—its mechanisms do not hide behind a skin or pretend to be anything other than what they are. Its technologically tentative construction appears to acknowledge the impermanence of objects in time, and it does not have the pretentiousness of futurism. In short, *K-456* presents a glimpse of what wabi-sabi might look like as technology: handmade, impermanent, and imperfect, choosing to get by with less and pursuing a quirky and flawed elegance.

Countertechnological Technologies: Perversions and Counter-oriented Identities

K-456 illustrates Stephen Duncombe's concept that DIY culture primarily exists as a negative and counter-oriented identity.[34] In other words, Duncombe argues that DIY projects are often countercultural and targeted to critique an established institution. Paik's *Robot K-456* has several targets, most notably the established framework of contemporary art that monumentalized the creative individuality of the modern artist and worked to build up a reverence around them. DIY-oriented movements like Fluxus were unified in their resistance to the dominant structure of modern art. This approach exemplifies Duncombe's view that DIY practitioners construct "who they are and what they do in opposition to the rest of society. Their identity is a negative one."[35] For this reason, instead of being an organized discipline, DIY artists are often a community of practice focused on opposing mainstream culture.[36] Most notably, *K-456* aggressively decommercializes the art object by challenging artistic conventions that physical objects should be kept, revered, and sold. Ironically, Paik's works have become some of the most valuable artwork of the twentieth century, which underscores how tactical DIY approaches often get swallowed up and absorbed into the institutions they are criticizing. The tradition of art, which has a long history of questioning its own institution and category, is often quick to co-opt critiques, which drive its annual churn. Nonetheless, everyday technologies and objects that are built in a rough DIY style invert our understanding of technology, and can similarly challenge our definitions of art.

This is a work of countertechnological technology, or to borrow Rafael Lozano-Hemmer's term, it "perverts technological correctness" by building an awkward implementation of technology to question the belief that advanced technology is a cure-all for social problems.[37] Paik's robot does not simply revel in the wonder of new technologies. It does not fall into the "effect effect" of simply amplifying the novel affordances of new technologies.

The convenience and futuristic qualities of Robby are swapped for a material honesty and a more complex, junky, and interesting thing. Technology is humanized through DIY by making it imperfect, and in the process it gives us a playful critique of social development.

Conclusion: DIY Practice as Wabi-sabi Antisolutionism

The point here is not to argue that Paik was deeply influenced by Japanese aesthetics and Zen Buddhism. Instead, the concept of wabi-sabi is introduced as a framework to better understand the attributes of do-it-yourself production in general—which also applies to Paik's *K-456*. The argument here is that junky DIY work is often more than just junk: the asymmetry, tentativeness, and roughness of objects can work to make things more interesting if we set aside the default "Robby the Robot" mode that manufactured goods typically pursue. *Robot K-456* represents a humanity and world that is imperfect, impermanent, and incomplete. Things break down, piles of junk accumulate, and the new exists alongside the old. The mindset of wabi-sabi is more of a garage sale versus a Costco, more hand carved than injection molded.[38] It embraces the beautiful mistakes of life.

The lesson we can pull out of *K-456* is that DIY practices often embrace a gritty reality that welcomes mistakes, plays with the residual layers of history inside outdated objects, and finds elegance in imperfections. This is a rarity in most consumer electronics: they typically function in a logical way until they break instead of functioning with quirky personality and building up character over time. In other words, mass-produced technological devices rarely age well; they typically stop working before they patina. As a result, DIY electronics and *K-456* are not a triumphant monument to flawless humanity. *K-456* presents an image of how wabi-sabi and countersolutionism takes form through DIY technologies, celebrating the handmade, impermanent, and flawed, which intentionally chooses to get by with less and strives for an unexpected and imperfect beauty.

4.2 Disobedience and *Hairbrain 2000*: Burlesque Technologies, Highlowness, and Neoretroism

Parody is a core technique of cultural production. Throughout this book I have argued that do-it-yourself producers often deliberately seek to create a nonideal version of professionally built technical artifacts. This style of rough-and-dirty recreation of something professional is often done for the sake of humor, especially when built from trash.

An example is found in Vũ Văn Nam, a seventeen-year-old high school student in northeast Vietnam that started building cardboard supercars with his friends using old bicycles as a drivetrain (figure 4.2.1). His first car was a Lamborghini Aventador SVJ, built using recycled cardboard and

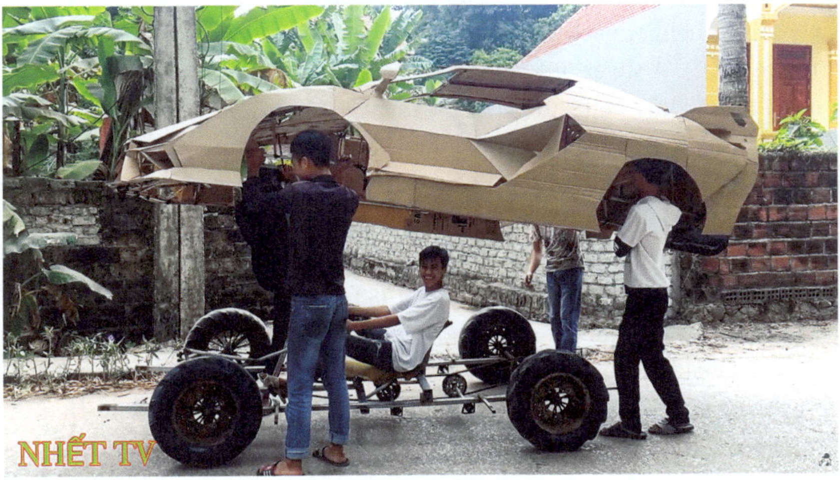

Figure 4.2.1
Vũ Văn Nam with friends working on their *Lamborghini Aventador SVJ* made out of cardboard. Courtesy of Vũ Văn Nam.

old bicycles—and the end result was a spectacular testimony to the stubbornness of humans despite a lack of resources. Although Vũ Văn Nam hadn't intended on his work being a humorous parody, it embodied how an eager group of individuals can independently reproduce a luxury item with nearly no resources. It clearly lacks almost all the qualities of an actual Lamborghini, but Văn Nam's work comes as close as possible with no budget. It is captivating work because it attempts to spin trash into gold like the old-fashioned fable Rumpelstiltskin.[1]

This form of imitating upper-class items with a lower-class budget can take many forms. For example, Thomas Thwaites's *Toaster* comes to mind, although Thwaites's work does not aim to reproduce a visual replica. To more clearly understand how artistic parodies work, it is useful to look at an example from Canadian artist Laura Kikauka. Her work intentionally blends junk with high-tech in an enlightening yet burlesque way that helps us more clearly understand the DIY mindset.

Low-Burlesque Technology and Laura Kikauka's *Hairbrain 2000*

Hairbrain 2000, by Canadian artist Laura Kikauka, is a prime example of how lowbrow DIY art can parody high technology. From a distance, the core of Kikauka's 1993 system looks like a set of virtual reality goggles with a complex helmet packed with electronic components and exposed wiring (figure 4.2.2). A silver full-body suit accompanies the helmet and goggles, and a heavy-duty belt slung around the waist holds packs of batteries. The helmet is initially quite intimidating; its packed components are placed consistently around the full crown of the head, resembling a scientific device that might be able to sense brainwaves, or a vintage hooded hair dryer from a 1960s salon. From a distance, the helmet is reminiscent of a science fiction movie like Trumbull's 1983 *Brainstorm* and features a head-mounted device capable of recording experiences.[2] It is so heavy that it requires dual shoulder braces to support the weight of the equipment worn on the head. When looking closer, however, it becomes apparent that the system is a high-tech facade cobbled out of relatively mundane items; the helmet is a metal kitchen colander with electrical relays stuck to its exterior, and the "VR goggles" are made up of a welding mask with a tin box added in front of the eyes. The entire contraption is not metal, only doused in silver spray paint. The body suit is not high tech, it is sprayed with expanding

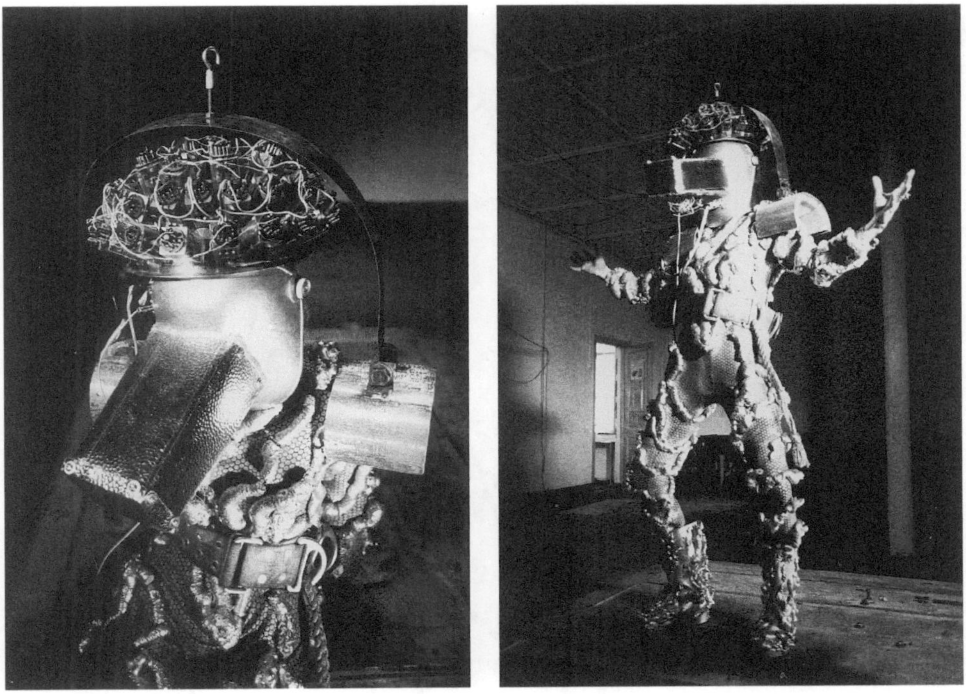

Figure 4.2.2
Laura Kikauka's *Hairbrain 2000* (1993), featuring a life-sized metal harness, a kitchen colander, a custom electric circuit triggered by ball-bearings, 48 electromagnets, and a battery pack. Courtesy Laura Kikauka.

foam from a hardware store. When closely inspected, it is more reminiscent of a prop from a B movie by Ed Wood than a Steven Spielberg blockbuster. Kikauka's clever construction, however, is absolutely intentional and much more than its external appearance.

For those wearing the contraption, the inside presents a completely different view of a functional, analog, and electromechanical piece of artwork (figure 4.2.3). The sheet metal box that extends out in front of the eyes functions like a kaleidoscope. The enclosure has mirrored internal walls and a custom-built circuit board on the bottom. Metal ball bearings wrapped in steel wool freely roll on top of the circuit board. When powered on, they cause electrical sparks to light up inside the mirrored vision chamber, which creates a geometric landscape of reflected light. In addition to visuals, the sparks activate an immersive field of audio on top of the wearer's head.

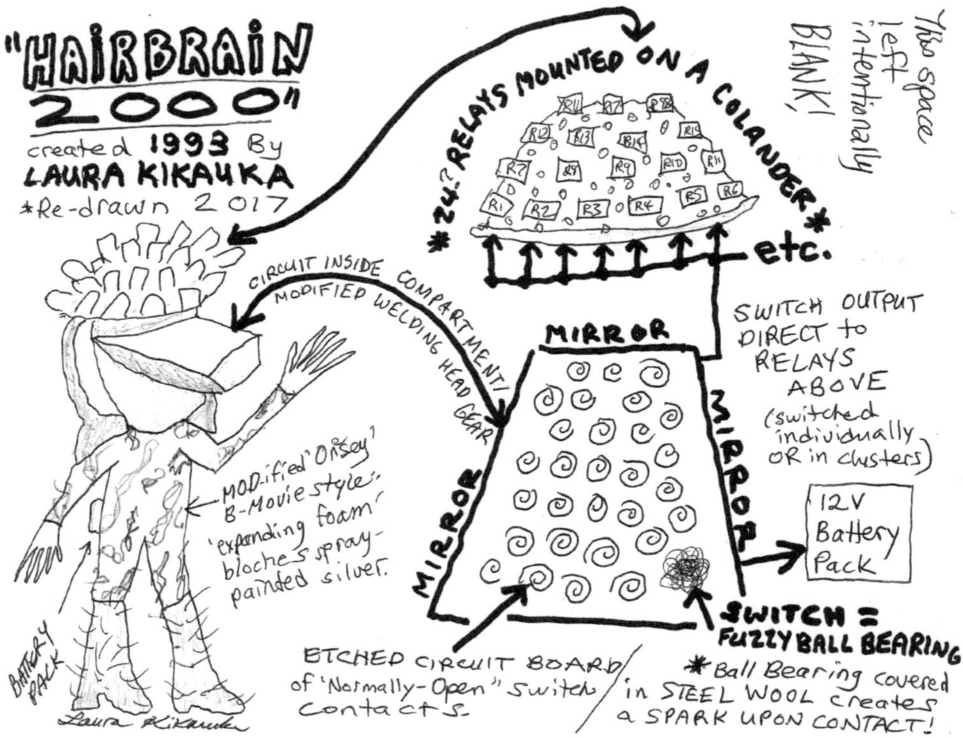

Figure 4.2.3
Laura Kikauka's system diagram sketch of *Hairbrain 2000,* drawn in 2017. Courtesy Laura Kikauka.

They activate electromechanical relays mounted on the kitchen strainer, which make a heavy clunking sound. When the wearer tilts their head forward, the ball bearings roll to the front of the mirrored chamber, creating sparks that set off the relays to fire an auditory barrage. Titling one's head to the right or left proportionally repositions the audio to emanate out of the right or left of the colander helmet, essentially operating as a spatialized, multi-channel audio system of clunking electromagnets.

Hairbrain 2000 parodies the usual refined and futuristic presentation of virtual reality. When speaking about her system, Kikauka says "I really loved the idea of using low tech to do things that were high tech. It was a bit of a parody of VR and how people look like when they're in it and how awkward it looks from the outside: it's only for the person in it having the experience."[3] The awkwardness between the "inside" and "outside" of VR became even more apparent when Kikauka took the system outside of a gallery environment and fumbled her way through the real world while wearing it (figure 4.2.4). She rode the subway and navigated the city, groping awkwardly around in her odd technology setup around confused onlookers (figure 4.2.5). Is this the future, or is it junk? Is this person brilliant or crazy? Technology often mediates communication and can make these distinctions unclear.

What we might call "burlesque" technologies are a clear attempt to discredit advanced technologies by making a low-budget parody of them. As a term, burlesque has considerably more use than only to describe old-timey strip shows: "burlesque" in its fuller sense describes when a low-class thing takes a swipe at a high-class thing—it originally describes class-based conflict through humor. As a result, a wider definition and analysis of "burlesque" is incredibly useful when attempting to dissect the dynamics of the DIY mindset.

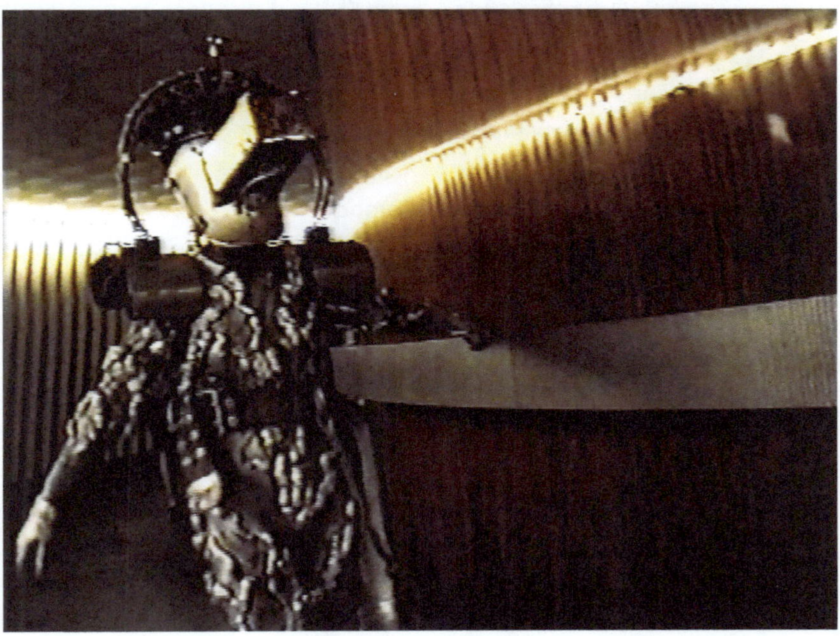

Figure 4.2.4
Video still from Laura Kikauka's *Hairbrain 2000* as she attempts to navigate the city while wearing the project. Courtesy Laura Kikauka.

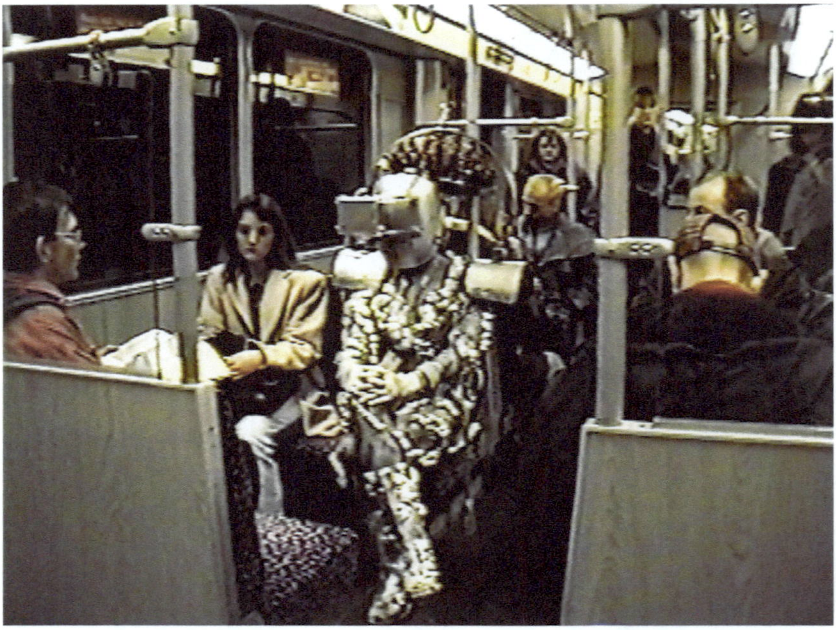

Figure 4.2.5
Video still from Laura Kikauka's *Hairbrain 2000* as she takes the subway while wearing the project in public. Courtesy Laura Kikauka.

Inside/Outside VR and the Sword of Damocles

To understand why Kikauka was fascinated by this disjunction between the interior and exterior, it helps to look at the origins of computer-based virtual reality, and its impact on popular culture and technical fields like ubiquitous computing. By being attentive to history, we can better track the persistence of themes like the stark difference between the inside and outside of VR.

The humorous disconnect between the awkward electromechanical system of VR and the futuristic promise of its rendered interior has always been present. Historians consider the first head-mounted stereo VR-type display to be what was called the "Head Mounted Display" system, created in 1968 by Ivan Sutherland's team. Sutherland's head-mounted display consisted of a metal halo that the user physically secured their head into. Cathode ray tube (CRT) displays, mounted like handlebars, swept back out of the wearer's eyes (figure 4.2.6 [a]). Sutherland's head-mounted display (HMD) used half-silvered mirror prisms in front of the eyes to insert wireframe three-dimensional objects into real scenes—an augmented form of reality where "displayed material can be made either to hang disembodied in space or to coincide with maps, desk tops, walls, or the keys of a typewriter."[4] Although other head-mounted stereoscopic displays had been used before the Head Mounted Display, Sutherland and his team were the first to use them with computer-generated images.[5] As background context, in 1963, while completing his doctoral thesis, Sutherland built the Sketchpad system, which is widely considered to be the groundbreaking predecessor to contemporary computer graphical user interfaces (GUI) and computer-aided design (CAD). In many senses, Sutherland was responsible for establishing many of the core paradigms for contemporary human–computer interaction (HCI).[6]

From the very origins of VR, stereoscopic, head-mounted displays have been physically awkward. Hence, Sutherland had named part of his HMD the "Sword of Damocles," implying that the system was so imposing that it resembled a death trap (figure 4.2.6 [b]). The Roman story of Damocles was about a fourth-century BC courtier who fawningly praised king Dionysius II of Syracuse for his good fortune and authority.[7] The king, wanting to convey the continual danger that came along with his role, proposed switching roles with Damocles for a day. He then installed a large sword above his throne that Damocles now sat on, precariously suspending it by its pommel by only a single hair from the tail of a horse.

In the case of Sutherland's headset, the Damocles reference related more to its formidable ceiling-mounted appearance than any immediate physical danger.[8] The unwieldy system hung from the ceiling directly above the user with a long, vertical, overhead pole that resembled Damocles's hanging sword. Sutherland's most accurate method of calculating the wearer's head position was accomplished through "rather heavy and uncomfortable to use" mechanical linkages, which involved an additional six-by-six-foot cross-shaped armature that was affixed by universal joints to the user's head.[9]

By the time Kikauka built *Hairbrain 2000* a quarter of a century later, virtual reality headsets were still largely unavailable as consumer commodities. They remained industrial and research tools for medical and military purposes until the 1990s. The late 1980s and early 1990s, however, saw a drastic upswing in consumer interest in the technology. This period was marked by a sense that digital information technologies were enveloping society. For example, *Wired* magazine, launched in early 1993, was one of the first digital

(a)

(b)

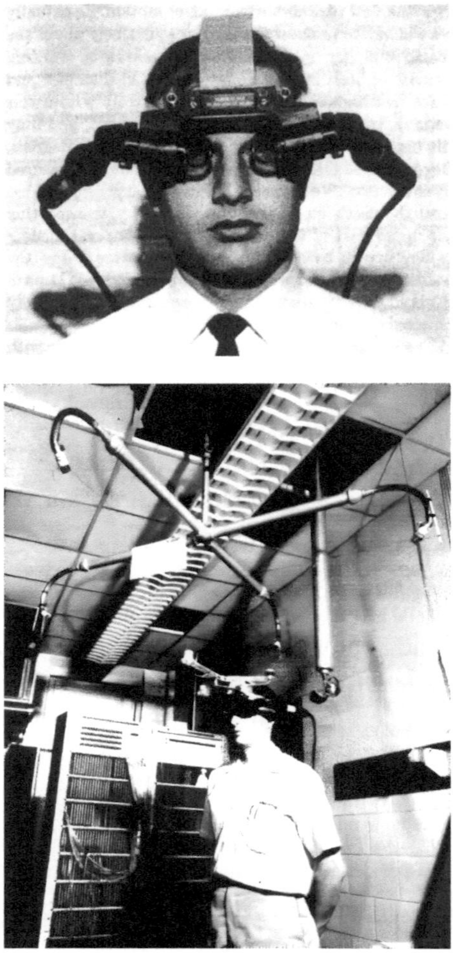

Figure 4.2.6
Sutherland's 1968 "Sword of Damocles" head-mounted virtual/augmented display system. Courtesy Donald Sutherland.

technology magazines written from the perspective of a lifestyle publication. Cofounder Louis Rossetto stated in the first issue that "the Digital Revolution is whipping through our lives like a Bengali Typhoon."[10] Through Hollywood movies like *Lawnmower Man* from 1992 and technology writers like Howard Rheingold, VR was promoted to the mass market as the next big thing.[11] Virtual reality was known widely enough by the general population in 1993 that Kikauka's piece would have clearly been recognized as a parody.

The excitement around virtual reality swelled up from a number of different directions before its peak circa 1993. A key catalyst was VPL Research, which was founded by VR enthusiast Jaron Lanier in 1984 and was the first commercial company targeted at making virtual reality available to a wider market. The company included the well-known electronic hacker Mitch Altman, who would later become a mainstay on the "maker" scene.[12] VPL developed the *DataGlove* hand-based interface system and the *EyePhone* VR headset. In many senses, the company was a major catalyst in spreading the futurology of VR into the public imagination. However, this excitement discarded the augmented approach pursued by Sutherland and others like Myron Krueger in exchange for being immersed in purely computer-generated worlds.[13]

Jaron Lanier started promoting the technology under the term "virtual reality" in 1987, and it soon became a technocultural obsession. Although several other designers and artists like Scott Fisher, Michael Naimark, and Brenda Laurel were working in the field, Lanier took on the mantle of public advocate for VR in the '80s and early '90s. Writing in 1993, David James Brown stated that the "diverse scope of possibilities created through full sensory immersion into computer-generated worlds caught the collective imagination, and Jaron became the hero of Cyberspace."[14] Lanier was an optimistic promoter, but generally stopped short of pushing VR as a kind of transcendental or technological panacea. The press and popular culture, on the other hand, were transfixed in the liberatory hype pf VR. For example, the front page of the *Wall Street Journal* on January 23, 1990, claimed that Lanier's work was "Electronic LSD," signaling the start of street-level excitement and hype around the technology.[15]

VR Skepticism and Ubiquitous Computing

The disjointedness between the interior and exterior of VR had many critics in addition to Kikauka. During the same period, a significant movement in computer science research had gelled in opposition to the "goggles and gloves" interface paradigm of virtual reality. The clearest and earliest objector was Mark Weiser, chief scientist at Xerox PARC, who published the influential article "The Computer for the 21st Century" in *Scientific American* in 1991. Weiser envisioned a post-desktop model for computing, a vision that heavily influenced the contemporary development of the "Internet of Things" (IoT) and charted a research agenda that led to the mobile, wearable technologies of today.[16] To him, the physical awkwardness of virtual reality provided a counterpoint to his idea of "ubiquitous computing":

> Perhaps most diametrically opposed to our vision is the notion of "virtual reality," which attempts to make a world inside the computer. Users don special goggles that project an artificial scene on their eyes; they wear gloves or even body suits that sense their motions and gestures so that they can move about and manipulate virtual objects. Although it may have its purpose in allowing people to explore realms otherwise inaccessible . . . virtual reality is only a map, not a territory. . . . Virtual reality focuses an enormous apparatus on simulating the world rather than on invisibly enhancing the world that already exists.[17]

In hindsight, Weiser's vision clearly outpaced Lanier's VR agenda. Many engaging VR artists emerged in the mid-1990s, like Canadian Char Davies. She had started developing VR projects in 1993, and released her landmark piece *Osmose* in 1995. After this wave, according to Mary Anne Moser, popular excitement around VR from the start of the 1990s subsided by 1996.[18] VR rose to stellar heights, then crashed and took two decades to reemerge, while ubiquitous computing plodded along a more gradual and multifaceted research path in academia and industry. Although VR systems are no longer ceiling mounted and have dramatically shrunk in size, the awkwardness of the user having their eyes occluded by a headset has remained consistent. For example, the Meta Quest 2 VR system has a twelve-page health and safety guide which includes warnings to stay away from balconies, stairs, windows, open flames, and sharp objects, and to not drive a bicycle or automobile while wearing the headset.[19]

On Highlowness and Neoretroism

DIY projects often revel in mashing up wildly disparate things, which I have referred to as "highlowness" or "neoretroism." Highlowness is the intentional mixing of technologies on two different vertical levels—high tech meets low technology. "Neoretro" refers to objects that span different chronological periods; such projects include Steampunk's mixing of Victorian or preelectric clockwork technologies with contemporary devices. Highlowness unites high tech and low tech in a single object, while neoretroism is an anachronistic futurism. Although different, both highlowness (vertical high and low tech) and neoretroism (a sideways past and future linking of disparate time periods) are somewhat interchangeable due to technologies correlating to chronologies, but the two terms are useful because low tech still exists in contemporary times. Both of the concepts of highlowness and neoretroism are "mash ups" or "jump cuts" between two mindsets—and both are commonly deployed as DIY practices.

A high-tech-meets-low-tech dynamic is rarely as dramatic as in Thwaites's *Toaster*, which was an attempt to make a modern appliance from rocks mined from the ground. It often can incorporate the old fashioned and the new, like Golan Levin's *NeoLucida* project—a Victorian-era camera lucida drawing aid updated and released on Kickstarter.[20] In some senses, it echoes the concept of "highlowness," although *NeoLucida* is more of a "neoretro" invention that connects a gap in time as opposed to Kikauka's bridging the gap of class.

Kikauka's project highlights the "lowness" of DIY practice as a theme. Kikauka and others leverage the gap between chosen materials and desired outcome as a form of "neanderthal in cyberspace" design challenge. This production technique is a recurring theme in do-it-yourself work, where builders tackle a design problem from a constrained and low-tech approach. The resulting entertainment value is perhaps similar to seeing the Flintstones building modern electronic conveniences carved out of stone and animated by animals.[21]

High and Low Technological Burlesque

The humor that results when lowbrow and highbrow meet can be described as burlesque, which is a common theme across different forms of DIY production. While "highlowness" is useful to think about the material approach of using low-tech materials to accomplish high-tech tasks, "burlesque" is useful to understand and describe the mechanics and dynamics of how high and low interact to construct a parody.

Burlesque, in the original sense of the term, has little to do with its contemporary usage that typically describes an old-fashioned striptease performance. As a *literary* term, burlesque refers to a particular type of parody where a serious or formal topic is intentionally treated in a lowbrow way. In the seventeenth and eighteenth centuries, "burlesque" could also refer to the inverse: lowbrow topics treated with absurd formality. A "low burlesque," for example, could be a slapstick or bawdy performance of Shakespeare, while "high burlesque" might be a banal activity described using

Shakespearean language. In other words, low burlesque takes the highbrow and pulls it down into lowbrow territory, while high burlesque takes the lowbrow and pulls it up into the world of highbrow formality. Both low-to-high and high-to-low forms of burlesque are satirical and actively mock formal conventions for the sake of humor and critique.

Kikauka's work functions as "low burlesque," where the promised future of a high-tech VR cyberspace is recreated with a welding mask, ball bearings, and silver spray paint. Writing about the project in 1996, Rafael Lozano-Hemmer described it as "perverting technological correctness." It brought the optimism and hype around virtual reality down to earth and questioned the popular assumption that new technology will produce limitless creativity and erase intolerance on the basis of gender, class, and race.[22] Burlesque, used in this way, is a perversion of the neoliberalist assumption that information technologies emancipate us from political and geographical limitations.[23]

According to UC Berkeley's Paul Duguid, two things that new technologies tend to promise are *supersession* and *liberation*. We are told they will supersede and replace the old by performing better and they will liberate us by delivering freedom. Both of these points—that new stuff will replace everything old and that it will give you freedom—is at the heart of how all new technologies are sold.

An important aspect of DIY low-burlesque work is that it often brings things down to earth by humorously illustrating the shortcomings of a new technology. Paul Duguid, a scholar of information systems, reminds us that new technologies *constantly* overpromise that they will liberate us from constraints:

> The first is the notion of supersession—the idea that each new technological type vanquishes or subsumes its predecessors: "This will kill that". . . . The second is the claim of liberation, the argument or assumption that the pursuit of new information technologies is simultaneously a righteous pursuit of liberty. Liberationists hold, as another much-quoted aphorism has it, that "information wants to be free" and new technology is going to free it.[24]

Supersession and liberation both downplay the social and physical complexities of technology. In the case of information systems, Duguid illustrates that these tropes persist due to the flawed assumption that information can exist without a technological medium. Duguid's tropes remind us that new technologies consistently overpromise and underdeliver; freedom from practical constraints should be interpreted more as a human desire than a material fact. As a low-burlesque parody, *Hairbrain 2000* plays with the practical awkwardness of strapping wired VR goggles to one's head and navigating around the real world. Kikauka's work clearly pokes fun at the promises that virtual reality will supersede anything or liberate us from anything. It cleverly mocks the user experience (UX) awkwardness of a head-mounted display by putting it into the real world. Artistic DIY projects like Kikauka's *Hairbrain 2000* pervert supersession and liberation through technology by reintroducing us to the negative components of advanced technologies through low-burlesque humor.

Conclusion: DIY Practice as an Exploration in Low-Burlesque Technologies

Two themes have been consistent in the history of virtual reality: unbounded optimism over the potentials of the system and the physical awkwardness of actually using the technology. These two variables have been present, almost like a yin and yang, since the first VR experiments by Ivan Sutherland and his team in 1968. In other words, the gap between promise and reality is a consistent trope in VR that spans over a half century. I would argue that it is also a trope in new technologies in general, and that DIY practitioners, artists, makers, and hackers can play a valuable social role by playfully transgressing and critiquing new technologies. Characterizing artwork like Kikauka's *Hairbrain 2000* only as a slapstick gag misses the larger point. Providing a counterpoint to the promised future of technology should be the agenda, and DIY practitioners do this particularly well.

Building a VR system out of ball bearings, steel wool, a welding mask, mirrors, a metal strainer, and some electrical relays shows a raw innovativeness. It actively pokes fun at VR, which has been struggling with many of the same physical user interface issues for half a century. The problems with the gap between the simulated world and the real world—even just the awkwardness of VR goggles or the motion sickness of participants—were present in Sutherland's *Sword of Damocles* in 1968 and persist today in platforms like the *Meta Quest 2*.[25] For this reason, *Hairbrain 2000* is as valid as ever, and is a worthwhile reminder of the consistent overpromise of technology in popular culture.

A core takewaway is to understand that DIY work often operates culturally as low burlesque critique of high culture. Looking back at the literary term burlesque and applying it to technological development helps us understand how relatively lowbrow DIY work serves a cultural purpose. Electronic artists using DIY techniques can play a powerful and poetic role in cutting through the marketing tropes that saturate new technologies. Subtracting themes like supersession and liberation as identified by Duguid leaves us with practical constraints. And playing with the awkward, physical, real-world constraints of VR is exactly what makes *Hairbrain 2000* humorous and useful as a piece of critically engaged making.

4.3 Disobedience and *Feral Robotic Dogs*: Hardware Activism, Communities, and Planned Obsolescence

In the summer of 2003, a dozen New York teens at the Bronx River Art Center participated in a DIY-style electronics workshop where they took robotic dog toys and transformed them into mobile pollution sensors. This three-month class was led by Artist in Residence Natalie Jeremijenko with the objective of highlighting the nature and extent of contamination present at the Bronx River. Manufacturing plants along the river were used in the 1800s and mid-1900s to convert coal and oil into gas for heating, lighting, and cooking, and these leached contaminants persist today in contaminated soil through modern-day New York City.[1]

The toys were not just modified in order to explore their technological interiors like Reed Ghazala's circuit bent instruments, they were deliberately augmented in a politically targeted way. Jeremijenko and her collaborators gave the students instruction in basic electronics and how to upgrade toy devices with pollution sensors usually used for alarms, breath alcohol checkers, and air quality control systems.[2] These gas sensors detected volatile organic solvents and polycyclic aromatic hydrocarbons—specifically the type of pollution present in coal and oil refining.[3] The sensors were linked with the motors to enable them to actively move toward areas that had dangerous levels of pollution.

In addition to adding pollution sensors, the toys were also augmented to navigate in the rugged outdoors by adding off-road wheels and stronger motors. The result were devices that looked like Mad Max versions of the Sony Aibo robotic toy dog (figure 4.3.1). When released on site, the modified devices swarmed across the landscape to lead the team to areas of high toxicity (figure 4.3.2). The devices were used as political tools—almost click-bait for the media—to highlight an environmental issue in a creative and

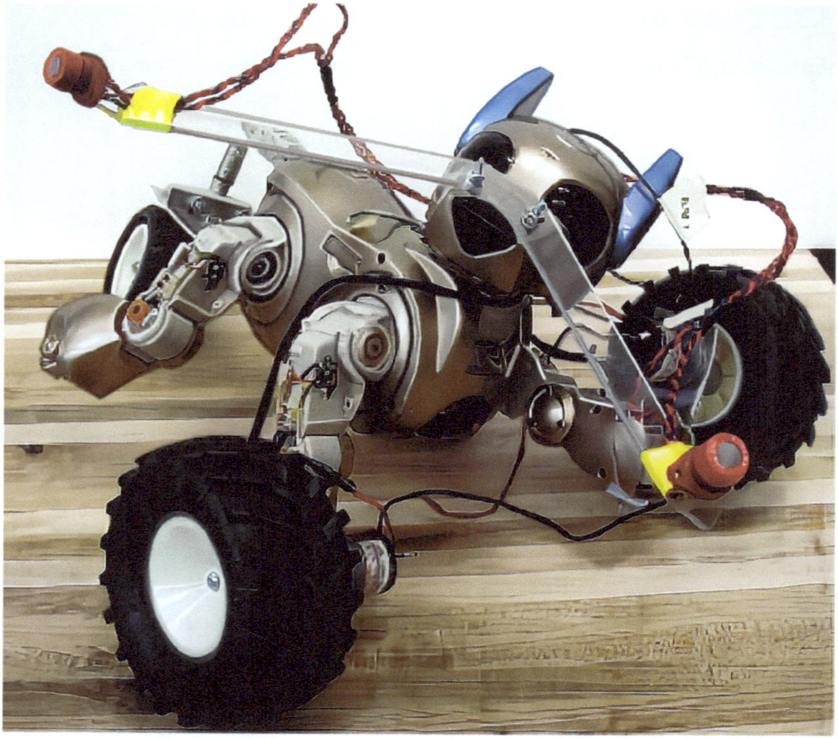

Figure 4.3.1
Feral Robotic Dog, built by Mike Kai following instructions by Natalie Jeremijenko and Will Kavesh. This device is a modified i-Cybie toy by Tiger Electronics, and wheels with motors have been affixed to the device to help it travel over rugged outdoor terrain. Volatile organic compound pollution sensors are mounted on a beam connected to the nose of the toy. Courtesy Natalie Jeremijenko.

publicly legible way. Jeremijenko hoped that the project would facilitate community-led monitoring of local environmental conditions and provide the opportunity for evidence-driven environmental discussion.[4]

When activated, the pack of robotic dogs can publicly sniff out environmental contaminants and move toward areas of high pollution. That this swarm-like activity is legible to a broad audience marks it as an activist tactic. A moving physical device in actual space is more compelling than a table of statistics or a bar graph. "The value of this project comes from who is doing the interpretation, who is doing the monitoring, and who cares," says Jeremijenko, "And when you release a pack of hot-rodded robotic dogs

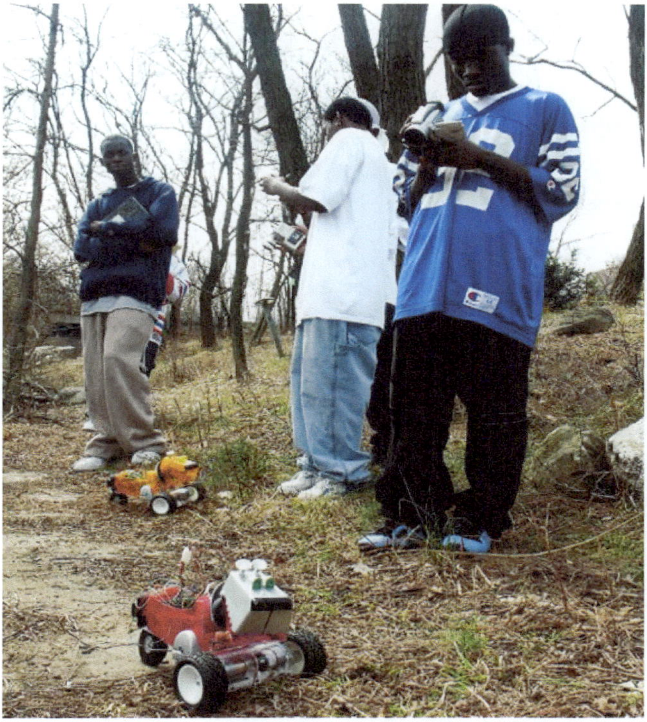

Figure 4.3.2
Students with *Feral Robotic Dogs* explore former gas plant locations for contaminants in 2003 in Starlight Park on the Bronx River in New York, NY. Courtesy Natalie Jeremijenko.

on a contaminated site, it becomes newsworthy."[5] *Feral Robotic Dogs* is a mediagenic public intervention, and it reiterates Rita Raley's concept of tactical media, which "focuses on open-ended questions rather than prepackaged lessons, instructions rather than products."[6] In this way, it can be thought of as being more *tactical* instead of *strategic*. In Michel de Certeau's terms, tactics and strategies are considerably different: strategies are activities that happen in a space defined by a clear sense of boundary, while tactics transgress across borders in a makeshift, guerrilla-style infiltration.[7]

Unlocking Activism through Technological Reappropriation

Jeremijenko started the project in collaboration with her students from Yale University in 2002. Enthusiasm for *Feral Robotic Dogs* spread through

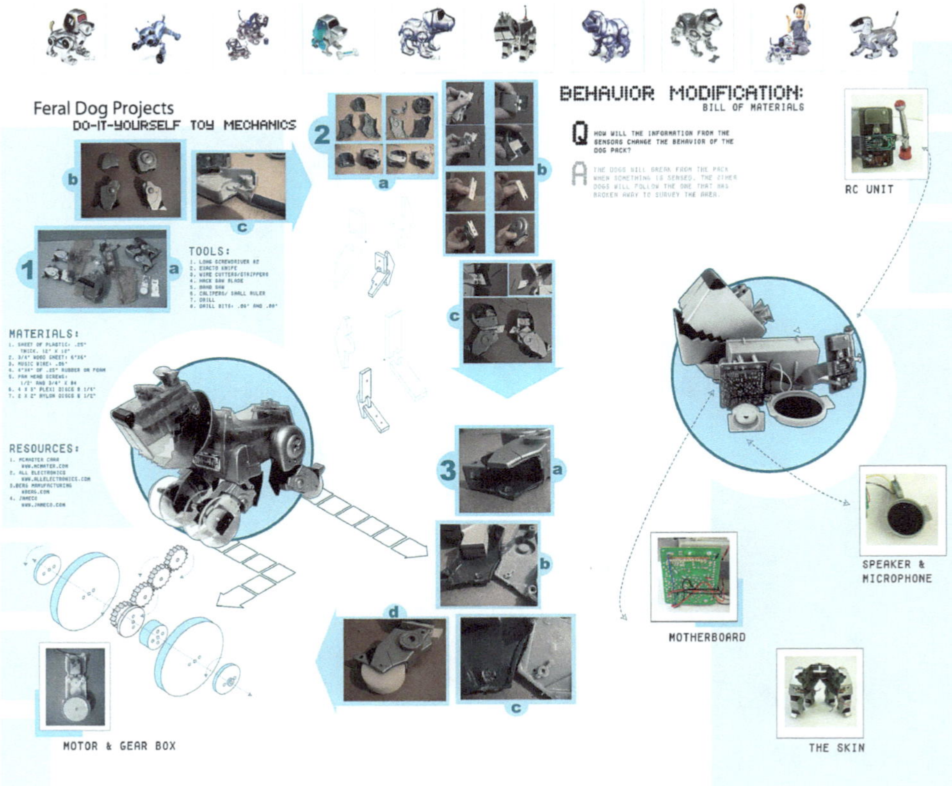

Figure 4.3.3

Feral Robotic Dog DIY instructions, including a bill of materials and tools needed for behavior modification with resources for the project. Courtesy Natalie Jeremijenko.

an assortment of public workshops where participants were taught to create their own devices. Reminiscent of low-cost DIY approaches like jugaad, robotic toy dogs were a platform chosen for economic reasons. They are an inexpensive collection of motors and sensors that if bought separately would be significantly more expensive. In this way, Jeremijenko's robotic dogs leverage the economic advantages of what Norm White would refer to as the "trailing edge" of technology: recently obsolete consumer technologies that are available at little to no cost.[8]

There is no specific device (or pack of dog devices) that is the "real" *Feral Robotic Dog*. Rather, the project exists as an open-ended prompt. "Turning technology feral" is more an artistic practice done in a group than a single completed artifact. The method of taking domestic toys and reviving them into tactical and "feral" devices follows four basic steps: getting a toy dog, adding rugged wheels, adding pollution sensors, and then changing the control logic. First, for the *Feral Robotic Dog* workshops, a battery-powered toy dog is supplied by Jeremijenko—one of the seven toy models that she and her team developed "reverse engineering" workshop instructions for. This repurposing of obsolete technologies echoes an overarching theme within the DIY mindset. This practice of feeding off of low-cost materials and obsolete technologies goes by many names—as zombie media practice (Hertz and Parikka), hacking the trailing edge of technology (White), circuit bending (Ghazala), or as "neoretroism" (Hertz).

Second, the toy is mechanically adapted to be able to move through an outdoor environment. The legs, which generally only work well on smooth indoor floors, are augmented with off-road-style wheels to better navigate rough outdoor terrain (figure 4.3.3). In some implementations of the project, participants completely remove the plastic legs and add axles below the dogs' torsos, while in other versions of the workshop, participants keep the legs and add wheels at the end of the feet. Third, air pollution sensors are attached to the exterior of the toy to enable the sensing of trace amounts of chemicals in the device's immediate environment. The dogs are equipped with a pair of sensors, which act like pollution-sniffing insect antennae that enable the system to detect whether higher pollution levels exist to the right or left of the device. Finally, the logic of the system is modified and upgraded to link the motion and activity of the toy robot to respond to the sensor data. The modified dogs are configured to smell out environmental contaminants and physically move toward these areas of higher pollution. Robotic dogs can be set up to communicate with other dog devices, record pollution levels, or perform individual behaviors when they find highly polluted sites: barking, playing the American national anthem, or rolling over and playing dead.

Citizen Science and Community-Driven Activism

Feral Robotic Dogs strives to empower groups who feel like they do not have a substantial voice in their own neighborhoods.[9] Gail Nathan, executive director of the Bronx River Art Center, described how the outcomes of the workshop flow back to the students, who, he believes, "conclude the course with a better understanding of the type of chemical pollutants that affect our fresh water system, a familiarity with the range of sensors that can be used to measure the presence of toxic substances in the environment, and a greater sense of personal interest in their role as custodians of the environment."[10] The project intentionally mobilizes amateurs and the general public like many other tactical DIY projects. Tactical efforts are often collective efforts, as there is always a need for the differentiated skill base that develops through collaboration. As Critical Art Ensemble has noted, amateurs have the ability to recombine personal elements and act as catalysts for action, for "specialization does not predetermine action."[11]

The devices are built close to a public outdoor site that is believed to contain high levels of pollution. Jeremijenko helps local community members build their robotic dogs in workshops, usually sponsored by a gallery, educational institution, or community center. Jeremijenko's workshops introduce community members to the basics of electronics and physical construction, while simultaneously involving them in locating environmental problems in their immediate neighborhood. In addition to dissemination through community workshops, the methodology for constructing *Feral Robotic Dogs* also takes place through web pages on the Internet. These pages invite members from additional communities to participate in the project. The utility of the project exists not just in the warnings it produces, but in how it amplifies voice and presents a locally situated blueprint for similar efforts. The approach releases data in a "feral" way, creating a spectacle of sniffing out environmental pollution and helps communities come together around shared problems.

What can we learn from Natalie Jeremijenko's *Feral Robotic Dogs* project? She reuses inexpensive and discarded robotic dog toys as a platform to detect pollution in public parks, schoolyards, and local communities. In the process, toys are transformed from domestic entertainment devices into tools to intervene in public space. The "dogs" are modified from their original form and equipped with sensors, rugged wheels, and new behaviors. Instead of performing tricks, the reengineered toy dogs create a public performance by using their newly installed senses to sniff out environmental contaminants (figure 4.3.2). In this way, the project acts as a creative form of citizen science. The devices are set loose in public areas of suspected

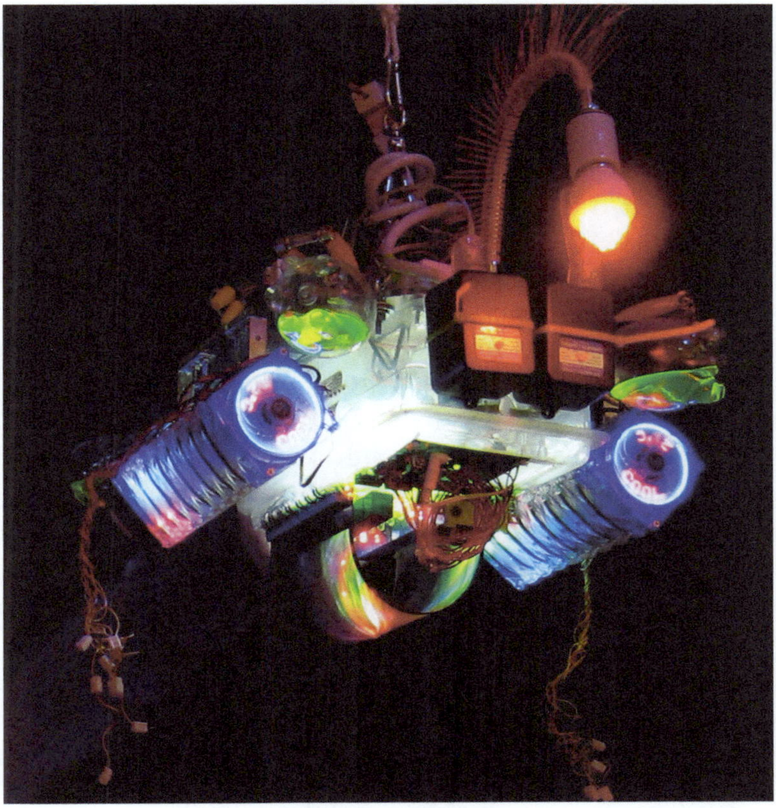

Figure 4.3.4
Detail of *Twilight Zone* by Shih Chieh Huang. The project repurposes consumer electronics in an unexpected way into a sculptural form. Courtesy Shih Chieh Huang.

pollution and move as a group toward high areas of contamination in order to create a photogenic spectacle that—in a style reminiscent of culture jamming—is intended to attract the media.

Activism against Planned Obsolescence

The millions of media technologies—including toys—that are discarded each year are an inescapable component to the ubiquity of computing technology. This substantial amount of electronic waste makes obsolescence a core dimension of today's technological landscape. In media theory scholarship, the utopian vision of the information age has a dystopic or heterotopic environmental underside that is largely neglected.[12] In other words, the obsolescence of information technology devices and how they are repurposed or end up as physical trash gets left out of most media histories, despite important steps made by some scholars, like Lisa Parks or Charles Acland.[13] In contrast to theory, Jeremijenko's work—along with a wide array of DIY electronic projects that repurpose outdated consumer electronics—can be seen as a practical and hands-on argument against planned obsolescence.

Other artists' work that clearly works against the planned obsolescence of electronics includes Katherine Moriwaki and Jonah Brucker-Cohen's Scrapyard Challenge Workshops, Ken Gregory's brilliant audio works, or Benjamin Gaulon's innovative "refunct media" systems that form long chains of repurposed information devices like a hacked obsolete media centipede.[14] Artists across the globe extensively repurpose obsolete electronics in their work, and it often produces more creative, textured and innovative results than only using newer technologies. Circuit bending, for example, joyfully ignores conventional opinions about what technologies are obsolete and what are useful.

As a background, the design of planned obsolescence was conceptualized by Bernard London in 1932, who saw it as a solution to the Great Depression. In London's mind, the economic downturn of the era was prolonged by consumers who continued to use and reuse devices long after they were purchased. His proposal was that every product should be labeled with an expiration date, and aftermarket use would be punished. "I propose that when a person continues to possess and use old clothing, automobiles, and buildings after they have passed their obsolescence date," he wrote, "he should be taxed for such continued use of what is legally 'dead.'"[15] Although London's proposal was never implemented, the concept of *planning and designing* obsolescence was widely adopted by product designers and commercial industry. Industrial designers like Brooks Stevens popularized the dynamic of planned obsolescence in 1954 as instilling a "desire to own something a little newer, a little better, a little sooner than is necessary."[16] Retail experts like Victor Lebow further clarified this mandate in 1955, saying, "These commodities and services must be offered to the consumer with a special urgency. We require not only 'forced draft' consumption, but 'expensive' consumption as well. We need things consumed, burned up, worn out, replaced, and discarded at an ever-increasing pace."[17] Technology hardware manufacturers wholeheartedly and intentionally adopt this approach.

Planned obsolescence takes many forms in contemporary consumer products: difficult to replace batteries in personal electronics and phones, proprietary cables and chargers that are only manufactured for a short period of time, software and hardware incompatibilities, discontinued customer support, or plastic enclosures that are glued shut and break if opened. Apple's AirPod headphones, for example, are manufactured with no user-serviceable parts inside, including their batteries. After a few years, the lithium polymer battery will no longer work, and the device will need to be either professionally serviced or discarded. The device is designed as a black box with no user-serviceable parts inside, and the product is specifically engineered *not* to be fixed. The advantages of this approach often result in significant gains, particularly for smaller and less expensive devices. However, this engineering comes at the high cost of often using raw resources that are difficult to source and producing harmful environmental contaminants when the devices are disposed of, often in countries with less economic development.[18]

Jeremijenko's project detours children's toys into a robust and localized social intervention. Discarded waste from mass production is exploited to criticize the very system of mass production that created it, by physically roaming around public space to sniff out hazardous waste. In other words, the *Dogs* are essentially smelling their own waste. Jeremijenko states that "most of the contaminated sites these dogs are exploring are the sniffing of their own butts, if you will, in a larger industrial ecology sense"—the project targets contaminants produced by the microprocessor industry and manufacturing.[19]

Jeremijenko speaks of her artwork as "feral," a reference to domestic animals that have escaped into the wild (*ferus* is Latin for "wild, untamed"). Like the circuit-bent work of Reed Ghazala that repurposes toys or the sculptures of Shih Chieh Huang built out of PC modification components (figure 4.3.4), Natalie Jeremijenko's work exploits economies of scale and surplus electronics to build work from an inexpensive commodity.[20] Mass-produced toys provide an inexpensive and readily available trailing-edge source of motors, electronics, and sensors for reappropriation. Jeremijenko combines the "hardware hack" mentality of circuit bending with a tactical sense of environmental activism. In the process, the project inverts the planned obsolescence of mass consumerism.

Expanding Tactics through DIY

Strategies belong to institutions, states, and economic powers. Tactics, on the other hand, belong to artists, activists, hijackers, and tricksters. Tactics do not hit on the frontline or target the center of power. Rather, tactical approaches slip through the cracks, launch an attack from an unexpected angle, disperse, and move on. Tactics are rhizomatic, while strategies are more centralized and arboreal. The *Feral Robotic Dogs* project utilizes the surplus of contemporary consumerism by "discovering holes in existent technologies and projecting potential change through these holes. Hackers call these holes 'exploits.'"[21]

As an artistic exploit of technology, the *Feral Robotic Dogs* project enriches our understanding of tactical DIY. It challenges institutional structures by repurposing media technologies as agitprop-style artistic activism that reuses technologies in a directly political manner. David Evans defines agitprop as artistic propaganda used for political agitation.[22] In this sense, *Feral Robotic Dogs* clearly operates as agitprop, with the attitude of tactical media, as a guerrilla attack on the institutions, powers of state, and corporations that have largely ignored public health concerns. Through the creation of open-ended, participatory "situations"—as opposed to works of art fixed in galleries—Jeremijenko's work hijacks consumer electronics as subversive critique. Because *Feral Robotic Dogs* creates a spectacle, its critique naturally attracts media attention. The *Feral Robotic Dogs* project also challenges narrow assumptions, preconceptions, and givens about the role older and outdated consumer electronics play in everyday life. In the process, Jeremijenko creatively activates a community using DIY methods through electronic art, visualizing environmental pollution in a poetic and memorable way, which helps us deal with the uncomfortable toxic realities of contemporary life in the twenty-first century.

Theme 5 Selling Out and *Graffitiwriter*

Co-option and DIY Electronics as Unpaid Industrial R&D

In 1920, American poet Ezra Pound wrote that "artists are the antenna of the race," a view shared by Marshall McLuhan and others who understood that artists regularly sense, shape, and sculpt potential futures. McLuhan, for example, saw artists as having an inside edge on understanding social changes, like a cultural radar system: "I think of art, as its most significant, as a DEW Line, a Distant Early Warning System that can always be relied on to tell the old culture what is beginning to happen to it."[1] But predicting or shaping the future often comes at a cost, with avant garde innovators most at risk of failure. Similarly, artists often face two different kind of enemies: the backlash of haters trying to stop you, and those that take your idea and sell it for profit behind your back. This second idea—of having DIY inventions taken behind your back and co-opted by industry—will be explored here. The topic of selling out forms an important topic within the DIY mindset, although when working with technologies, the thorny topic of authorship can be multilayered and complex.

One noteworthy art project that highlights the dynamics of DIY co-option is *Graffitiwriter*, a robotic system built in 1998 by a Pittsburgh-based group of university students calling themselves the "Institute for Applied Autonomy." The system, which resembles a hefty remote-controlled car, is a radio-operated device for remotely spray-painting text messages on the ground in the style of a dot matrix printer (figure 5.1). In other words, the four-wheeled robot squirts five spray cans downward to spell out blocky messages as it travels at approximately fifteen miles per hour.

The cans hang off the back of the robot like a rear bumper, with the row of cans perpendicular to the motion of the machine. The device is about the

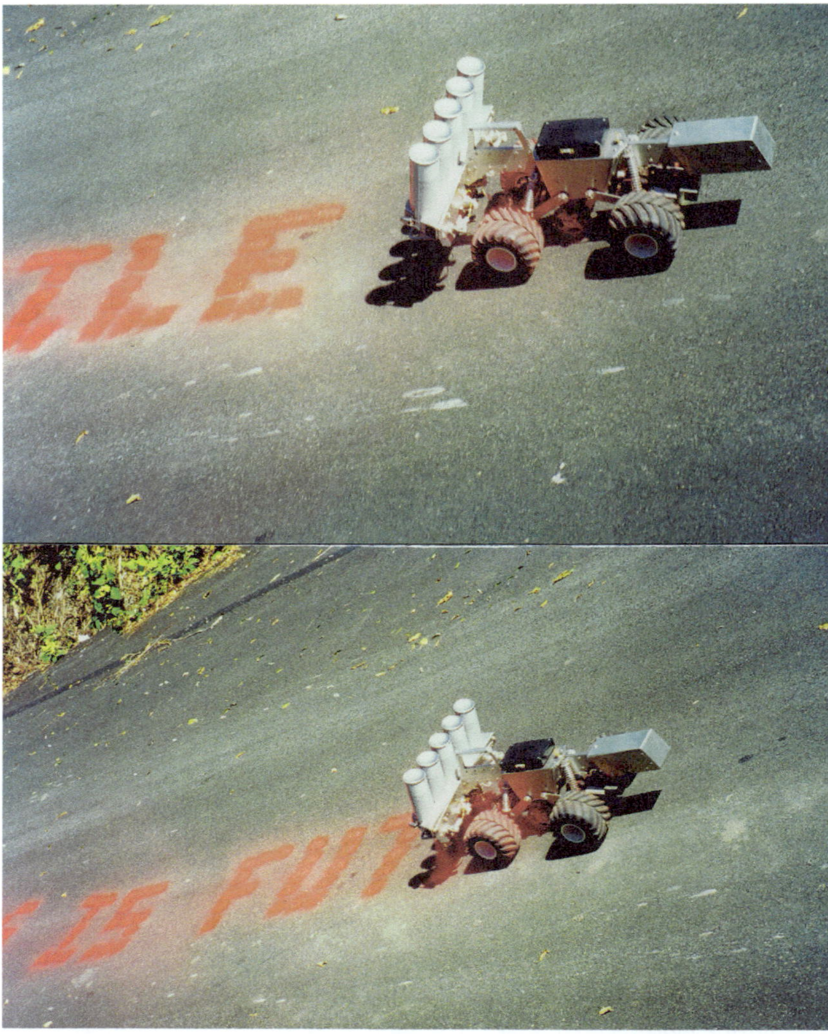

Figure 5.1
Graffitiwriter by Institute for Applied Autonomy is a robotic system that spray-paints textual messages on the ground. Courtesy Richard Pell.

size of a medium-sized dog, about 24 inches long, 18 inches wide, and has four off-road-style wheels. A black unit on top in the center where a truck cab would be holds the antenna. A manually machined aluminum frame contains a parallelogram-shaped blocky nose protruding at the front of the device that holds the batteries and counterbalances the spray cans hanging off the rear of the vehicle (figure 5.2). Inside the *Graffitiwriter* is a Parallax BASIC Stamp microprocessor with 16K of memory that stores sequences of solenoid patterns for each letter and the message to be written by the machine. A light-emitting diode and a photoresistor with a four-section pinwheel ring acts as a rotary encoder. This LED is attached to the robot's wheel axle and helps the system keep track of how fast the vehicle is moving, so it can write messages with letters that are the same size regardless of the speed the machine is traveling.

The *Graffitiwriter* as Tactical Media

A useful framework to think about Graffitiwriter is through the concept of *tactical media*. Tactical media proposes small-scale, quick-acting, and nomadic acts of resistance and disturbance as opposed to an organized political revolution. Tactical work engages in a micropolitics of disruption,

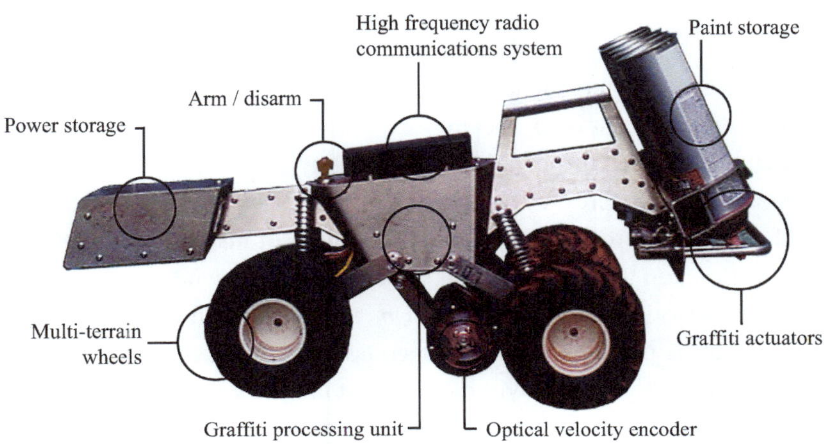

Figure 5.2
A diagram of the functional parts of *Graffitiwriter* including paint cans, multi-terrain wheels, and custom electronic systems. Courtesy Richard Pell.

intervention, and education. As a form of "culture jamming," it often "signifies the intervention and disruption of a dominant semiotic regime, the temporary creation of a situation in which signs, messages, and narratives are in play, and critical thinking becomes possible."[2] Tactical devices embody an activist argument or political action, and in the process remove personal inhibitions or social conditioning from the user. Information scholar Leah Lievrouw interprets the rise of tactical media as a response to the fall of communism in Europe in 1989, which also was accompanied by the failure of left-leaning revolutionary politics and the forward march of market capitalism. [3] Somewhat similarly, Raley interprets tactical media as a direct response to postindustrial society and neoliberal globalization.[4]

Graffitiwriter illustrates the current relevance of public space and protest in the political sphere. Lyn Lofland writes about "bubbles" of one realm that intrude into another that can "transform the character of a substantial portion of the space in which it is located."[5] *Graffitiwriter* extends this idea in a more radical and even illegal direction, semipermanently turning privatized areas into spaces for public communication. In this sense, it goes even further than a bubble. *Graffitiwriter* loudly proclaims that people have a "right to the city"—a concept promoted by sociologist Henri Lefebvre to push against capitalism's hunger for dominating urban space.

The Institute for Applied Autonomy and DIY-style Institutions

The choice for DIYers Richard Pell, Tad Hirsch, and the collaborators that built *Graffitiwriter* was to operate as the Institute for Applied Autonomy, which was inspired by operating in a legal gray area. They believed that the title would divert press coverage of the project away from the individuals and toward the project's critique of the systems of institutional knowledge production and funding. It was in some ways a DIY pseudoinstitution. IAA terms this work as *contestational design*, "a unique form of design activity whose aim is promote particular agendas in contested political arenas."[6] Just as robots deploy in environments that are physically dangerous, *Graffitiwriter* operates in spaces of political confinement, "in areas where free speech has been regulated out of existence."[7] *Graffitwriter* is like a robot that removes asbestos from aging pipes in a building or one that is deployed to help in a nuclear accident; however, instead of avoiding physical risk, *Graffitiwriter* helps avoid social and legal risk as an autonomous protest vehicle behind the façade of an institution. Extending the use of the idea of being contestational, IAA terms *Graffitiwriter* and other similar works *contestational robotics*, or "a research effort that inverts the traditional relationship between robots and authoritarian power structures by developing robots to meet the needs and budgets of culturally resistant forces."[8] These devices foster higher-risk forms free speech about government surveillance, public space, and law enforcement—and the official-looking robotic format enables the device to get away with illegal acts on the street.

The system initially wrote out messages that were programmed by the IAA, like "voting is futile." Later, the public was able to enter in any text, like "cool as hell" (figure 5.3). The machine had a relatively refined level of finish that "leveraged techno-fetishism to confer a kind of legitimacy to robot-mediated criminality."[9] In other words, *Graffitiwriter* was intentionally designed to appear sophisticated and official, despite its DIY construction (figure 5.4). "The assumption was that anyone possessing a robot represented a large research institution which probably had the right to spray

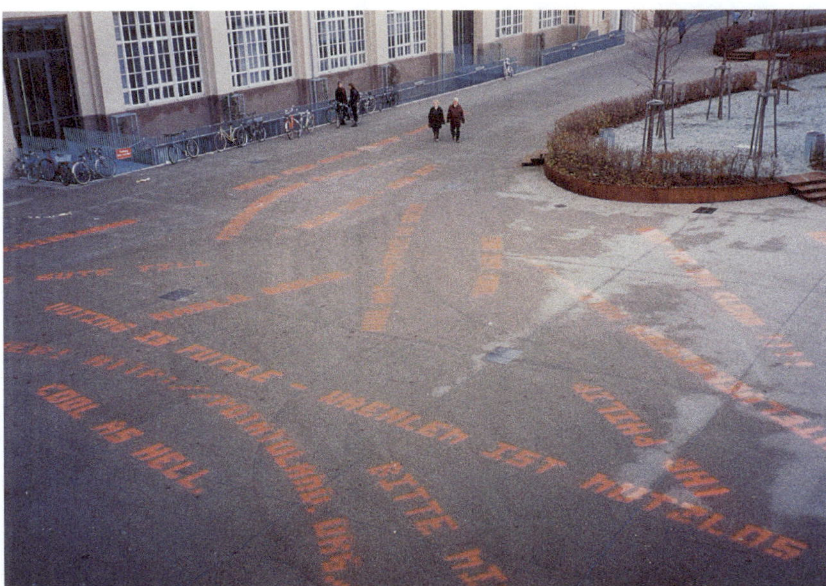

Figure 5.3
When operating in public, *Graffitiwriter* can quickly fill a large space with multiple messages. The system operates in Stuttgart, Germany outside the ZKM Center for Art and Media in November 1999 at the NetCondition event. Courtesy Richard Pell.

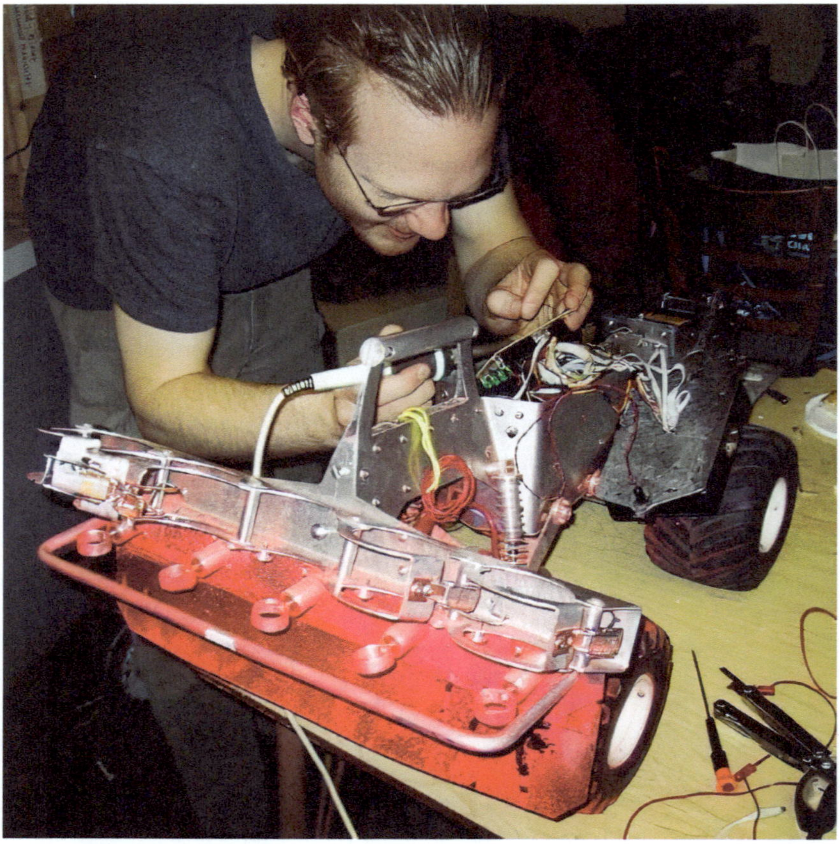

Figure 5.4
Graffitiwriter is a functional device that was engineered to be deployed in live, public space. A member of the Institute for Applied Autonomy works on *Graffitiwriter*'s radio communications system. Courtesy Richard Pell.

its messages on public space, rather than simply a couple of crazy people who built a machine in their garage."[10] As a research agenda, the Institute for Applied Autonomy asks why there are not robotics or research institutes that are designed to explore how humans can be more engaged, autonomous and democratic citizens.

The Institute for Applied Autonomy used the symbolic title of an "institute" to lampoon the idea of an academic research organization that pulls in research funding and outputs scientific knowledge. In other words, IAA is an "institute" like SRL is a "lab" or the Centre for Metahuman Exploration is a "center."[11] These DIY-style institutions lampoon formal institutions in general by adopting the facade of institutionality, but they regularly pervert standard research goals with anti-establishment "critical" logic. The knowledge output from IAA as an institute is not standardized knowledge like peer-reviewed academic papers in a scientific journal. This work consists of objects that get deployed through tactical performance more as a spectacle or a public intervention that lives on in documentation. These acts could be called "research-creation," a term that the Social Sciences and Humanities Research Council of Canada uses as a middle ground between critically engaged artwork and research. It is "an approach to research that combines creative and academic research practices, and supports the development of knowledge and innovation through artistic expression, scholarly investigation, and experimentation."[12]

Functionality is one way, for example, that *contestational design* differs in a significant way from *critical design*—critical design often does not construct functional technologies. Critical design projects often resemble prop prototypes that are primarily constructed to cause debate. Functionality is vital enough to this process that IAA's founding slogan was "our shit works"—the act of contestational design involves considerable engineering effort and function to be successfully deployed live as an intervention. IAA describes themselves as employing "methods of self-determination through artistic expression and application of military-like technology to the topics of criminal mischief, decentralized systems, and individual autonomy."[13] These measures are not standard electoral politics. Rather, the *Graffitiwriter* aligns to nonviolent civil resistance techniques like strikes, blockades, sit-ins, activism, or sabotage, and often involves some form of civil disobedience where laws are intentionally broken to make a political statement.[14]

Extending *Graffitiwriter*: *Streetwriter* and Asimov's First Law of Robotics

After the original *Graffitiwriter* was built by IAA, the team continued to develop a larger version called *Streetwriter* (figure 5.5). It was initially built into a cargo van in 2001, and in 2004 was moved into a trailer pulled by an automobile. *Streetwriter* had a similar computer-controlled spray-painting system to the original *Graffitiwriter* and was able to print text messages onto

(a)

(b)

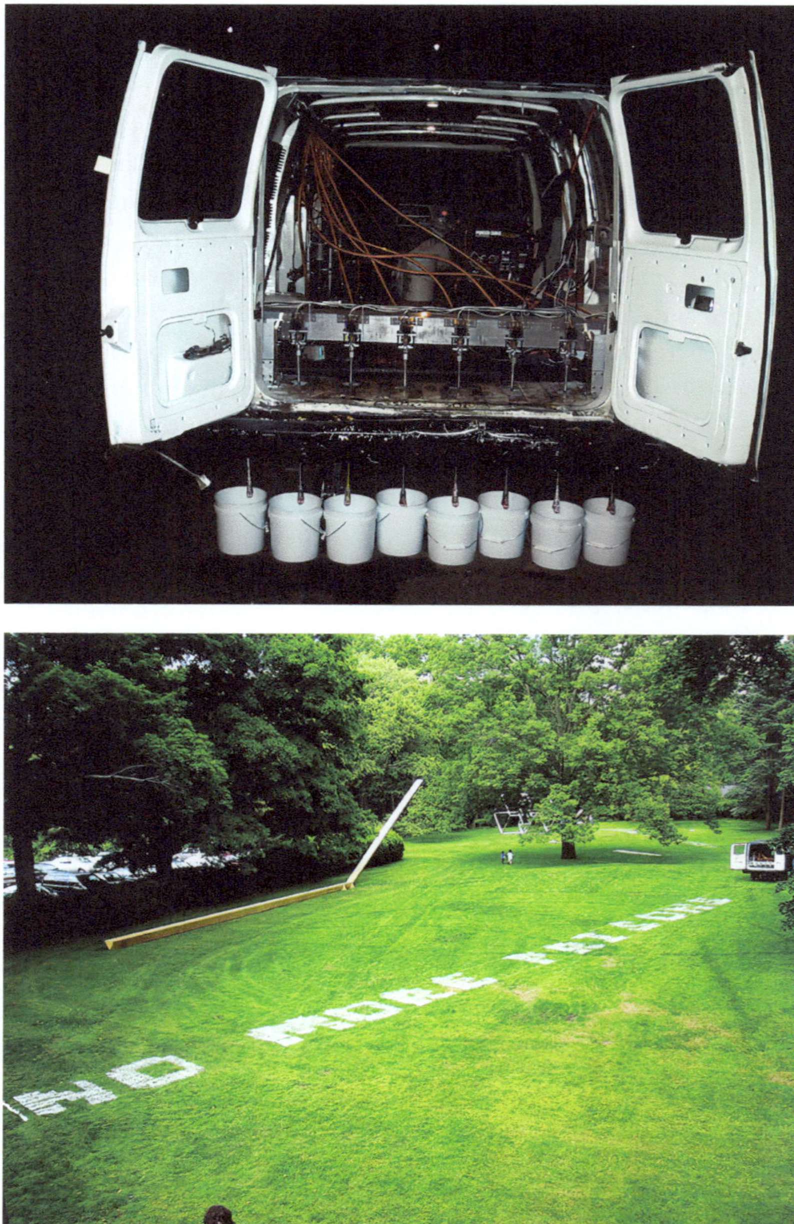

Figure 5.5

The Institute for Applied Autonomy's *Streetwriter,* which physically upscaled the *Graffitiwriter* sytem to be the size of an automobile, and enabled it to write larger messages. Courtesy Richard Pell.

the pavement like a dot matrix printer. The larger size allowed for jumbo-scale messages and graphics that were several hundred feet in length.[15]

As a capstone to the *Streetwriter* project, IAA used it to infiltrate the finish line festivities of the first DARPA Grand Challenge in 2004. The DARPA Grand Challenge is a high-profile competition for American autonomous vehicles funded by the Defense Advanced Research Projects Agency (DARPA), the leading research organization of the United States Department of Defense.[16] As a protest to the military-funded event, *Streetwriter* wrote the message "a robot must not kill" in large letters near the finish line for the event. The reference is to Isaac Asimov's Laws of Robotics, a set of principles outlined in his influential Robot series intended as a core framework to guide the behavior of autonomous robots that have decision-making abilities. This framework was first written in 1942 and includes two other laws: for the robot to obey humans if nobody is killed, and for the robot to protect its own existence if the other two laws are obeyed.[17] Asimov's laws have been central in helping articulate safeguards around robotics, communication technologies and artificial intelligence technologies in contemporary society.[18] *Streetwriter* used Asimov's laws of robotics as an unconventional and provocative appeal to the engineers, funders, and organizers: you are designing self-driving military vehicles that easily could be let loose to autonomously kill humans on their own with very little public discussion about whether this is a good idea.

Grafftiwriter and *Chalkbot*: A Co-option by Nike

Homebrewed artistic DIY projects regularly get taken by professionals in commercial creative industries and turned into products, advertisements, or platforms. In some instances this co-option of the amateur into the professional is completely normal and organic; at other times it is a blatant rip-off and theft of intellectual property.

This is widely referred to by sociologists and cultural studies researchers as "co-option," or when countercultural movements get picked up and used by the commercial mainstream, including Dick Hebdige, Thomas Frank, Joseph Heath and Andrew Potter.[19] Hebdige and Potter summarize the phenomenon as follows:

> First a subculture arises around a certain style of rebellious consumption, such as punk rock, skateboarding, or organic produce. In its original form, this subculture

> is genuinely subversive; that is, it poses a genuine threat to the established capitalist order. But as the subculture becomes more popular, corporations move in. They take the elements of the subculture, bleach out the subversive elements, and sell a denuded, non-threatening version of the subculture to the masses. This is known as co-option.[20]

The gravitational pull of co-option is a strong dynamic in all forms of DIY production. The pattern of DIY-style work getting co-opted into professional work is a well-worn path, especially for those working with custom-built technologies. In other words, many independent electronic and media artists know of someone whose noncommercial work has been taken, adapted and used commercially without authorization.

In 2009, five years after the Institute for Applied Autonomy used the system to protest autonomous military vehicles, an uncannily similar "paint-messages-on-the-road" device to IAA's *Streetwriter* appeared at the Tour de France. The device, called *Chalkbot*, was produced by advertising agency Wieden+Kennedy and built by the Pittsburgh design studio DeepLocal for Nike's "Livestrong" campaign.[21] With Nike's *Chalkbot*, one of the early *Graffitiwriter* team members had decided to covertly take the project concept and sell it to Nike via the advertising agency.[22]

The Institute for Applied Autonomy was especially upset that their anti-corporate project had been reappropriated into an advertising enterprise by a large and prominent multinational corporation. As a response to Nike, IAA issued a press release asking that the *Chalkbot* project do the following four things:

1. Publish their plans+code, in keeping with the open nature of the project.
2. Prominently feature a historical accounting of the technical and ideological origins of the robot on their website and in related publications.
3. Make the *Chalkbot* available for use by anti-corporate activists, free of charge.
4. Provide proportional financial support to new projects that share the antiauthoritarian and anti-commercial aims from which this project emerged.[23]

In summary, IAA wanted *Chalkbot* to remain open source, to have a history that acknowledged artists, to be physically shared, and to give financial support. The *Chalkbot* builders replied by stating that while they appreciated IAA and strongly resonated with its activist tendencies, "the *Chalkbot* lives on its own, however, as a product of four dedicated people who worked night and day."[24]

The *Chalkbot* engineer issued a short statement acknowledging the historical omission, but no acknowledgment of prior work was ever included by Nike. Although the specific details of *Graffitiwriter* vs. *Chalkbot* is not the larger point here, the main difficulty that IAA had was that the noncommercial DIY community was taken from and nothing was given back to it. In other words, IAA primarily wanted acknowledgment and for the commercially oriented device to be open sourced. In hindsight, these details provide a rough sketch about positive and negative ways for companies to use ideas that originate in noncommercial DIY environments. The way that *Chalkbot* used *Graffitiwriter*'s intellectual property is specifically an example of what *not* to do.

The core point here is part of a larger issue of co-option—"the corporate appropriation of 'outsider' research projects without acknowledgment of the amateur, collective, hobbyist, and activist communities upon which projects like *Chalkbot* are built."[25] Apart from the specific details of this case, *Chalkbot* reminds us that tactical, anti-consumerist, and subversive projects can easily be adapted into different commercial forms. In other words, a technology's use can be flipped by changing its context. This is both a blessing and a curse for DIY producers. For those wanting to do independent noncommercial work, it is a constant threat. For those wanting to professionalize themselves, it is an opportunity. For commercial professionals, the DIY electronics scene of artists and experimental designers is often a good source of fresh ideas.

Conclusions: DIY Electronics as Unpaid Industrial R&D

In conclusion, it is important to highlight that many of the most common and respected technologies were built by creative amateurs. Often this work is dismissed as whimsical or impractical only to be copied later by industry. High-profile research labs know that artists are useful—the MIT Media Lab, the Atari Research Lab, Xerox PARC, Bell Labs, and other historically significant facilities have all employed artists to aid their innovation processes.[26] In discussing the topic of artists in labs, media artist Golan Levin proposes that in these contexts artists aid in predicting the future: "To get a jump on the future, in other words, bring in some artists who have made theirs the problem of exploring the social implications and experiential possibilities of technology. What begins as an artistic and speculative experiment materializes, after much cultural digestion, as an inevitable tool or toy."[27] Or, in other words, artists and amateur technologists are often more adept at uncovering innovative user experiences and unexpected or playful social interactions than traditional research labs focused on engineering-led or marketing-led approaches. In his analysis, Levin proposes several notable examples: Myron Krueger's 1974 work was copied by the Sony EyeToy in 2003, Michael Naimark's work from 1978 was taken by Google Street View in 2007, Art+Com's 1996 work was copied by Google Earth in 2001, and Chris O'Shea's 2009 work was copied by Forever 21 in 2010.[28] In these cases, the work of independent artists served as unpaid industrial research and design.

Projects like *Graffitiwriter* and work by other culture jammers "could cynically be designated as an outer-edge and wholly unpaid research and development department for the global marketing industry"[29] by exploring new modes of public engagement in the global attention economy. Artists, experimental designers, and DIY practitioners regularly invent and innovate in ways that industry does not.

As a guide for DIY technologists, techniques to overcome unwanted co-option include attempting to document your work clearly in public to create your own history, taking your grievances about having your project stolen to the press, or placing your work under a creative commons license that pushes those that borrow it to also share alike. There is no single solution.

As a tip to advertising agencies and others in the commercial creative industries: please communicate with the DIY producers you are inspired by. Clearly and proactively talk to them and acknowledge them. In this case, the Institute of Applied Autonomy primarily wanted acknowledgment from Nike, not money, in terms of taking the idea and turning it into *Chalkbot*. Although financial reward likely would have been happily welcomed, Nike simply failing to publicly acknowledge the originality of a piece of DIY electronic art like *Graffitiwriter* was the core offense.

Conclusions The DIY Mindset

So what use are these cobbled-together electronic technologies in the face of all the problems facing our world today? Beside existing as creative technological curiosities, what do these examples have to teach us?

My answer to this question is that do-it-yourself methodologies are absolutely essential. It is vital to understand the ubiquitous, innovative, subcultural, and disruptive potentials of DIY practices. DIY is everywhere. In the end, DIY approaches are not cure-alls or catchalls, but are rather large and amorphous categories that deserve our attention from artistic, technological, and cultural perspectives. It is my hope that the concepts and terms presented in this book have helped start to fill out a more fulsome lexicon of DIY practices. Whether we describe them as disorienting, transparent, performative, disobedient, burlesque, co-opted, or something else, the core thing to remember is that DIY practices and the artists that create these technologies are important. They hold a key to both the past and the future: artists have clearly been into the "maker movement" for over a hundred years, and they will continue producing innovative work for several more centuries.

DIY Practice Forms the Core of How Contemporary Artists Use Technology

DIY methods are a kind of human-scale working process. DIY is a techno-vernacular mode of invention found across the globe that goes by many names—and its amateur roots are at the core of them all, whether we refer to it as jugaad, gambiarra, ad hocism, zizhu chuangxin, junkyard innovation, jua kali, or Système D.[1] DIY is a human-scale and nonindustrial way of quickly putting things together when few other options exist.

An important element to remember is that with DIY's ubiquity comes hardship. DIY practice in the real world is not sexy. It is most commonly borne out of a lack of resources. To think of DIY as only cool is to completely miss the economic reality of most of its origins. DIY sucks. It's hard. It is difficult. Although many examples in this book may look glamorous in this format, they generally grew out of a lack of resources.

In addition, an emphasis on DIY is an emphasis on the amateur. This is not to propose that we should do away with professionalism, but rather that we should pay attention to how nonscientists and nonengineers create things outside the standard brackets of science, engineering and commercial product design. Yes, we still need professional engineers to do stress analysis tests to make sure our bridges and buildings are safe, for example—but we also need to pay attention to technologists on the outer margins of industry. The historical record clearly shows that radically independent technology producers are a flourishing source of raw innovation. Pop and tech culture owes a great deal to its outsider and DIY practitioners since they are regularly its heartbeat of innovation.[2]

The Maker Movement Is Dead

Make magazine, for example, can be seen as a start to understanding the vast potential of amateur-oriented technology creation. But the maker movement is only the beginning. The fascination with cool tools—particularly Arduinos and 3D printers—was a diversion from the real superstar of the scene: curious and creative producers of technology that do not fit into an existing disciplinary category. To identify, cluster, and brand these folks as "makers" was truly a stroke of genius by Dale Dougherty, but the opportunity to do something actually interesting with the term was mostly squandered on promoting 3D printing. One key error was not staying focused on its core DIY community values. *Make* was not prepared for the complex challenges we face today—and as a source of ideas it is dead anyhow. We can safely proceed with or without it.

I'm interested in taking the term "maker" back and unbranding it by simply calling it "DIY." I also refer to it as "DIY electronic art," "the DIY mindset," "punk technologies," and "critical making." It is not that "maker" is insufficient, however—it is useful to explain to people that you fabricate things, or that you know how to use an Arduino and a 3D printer—but because of the frenetic emphasis that *Make* placed on the gadgetry of technology, the term is of little use today beyond that. The fad has passed, and it makes more sense to reflect on the fact that the maker movement wasn't a contemporary movement at all, but part of a century-long continuum within an experimental technological subculture. That subculture quietly thrives in communities scattered around the globe—as it has for about a century. Most of the contemporary scientific and art worlds have ignored this genre of artists-as-inventors, unfortunately, and a staggering amount of historical work needs to be done in order to expand our knowledge of historical precedents. It is massively interesting, varied, and uncharted: I will help you as best I can if you want to research this underexplored field.

The core part of the story of how artists work as technologists is still to be told. This book is meant as a start, and the themes and examples discussed here are only the outlines of a larger rubric. The categories all overlap with each other anyhow. As a part of this, I invite you to help me build this future together by contacting me and getting involved in helping craft the face of the "post-maker" movement.

DIY Practice Is Absolutely Essential in Carving Out Subcultural Identities

It is worth remembering that DIY practices form a core component of subcultural identities. Whether queer, Black, a skater, or an artist—or all of the above at once—DIY and independent cultural forms are essential. By focusing on DIY approaches, we highlight many important voices from the margins. The marginalia are actually much more interesting than the drab main text anyhow. Life flows at its edges.

The world of electronics is only one of many where independent work flourishes. Peeling away a layer of professionals in any field reveals a dynamic underlayer of producers. It is a bit like turning over a large rock: the surface looks stable, unified, and solid, but underneath is a complex and mixed network of crawling insects. The rocks of industry—the iPhones, the Stratasys 3D printers, or other professionally built technologies—give no indications of their interiors and have few clues about how they were made. On the other hand, the crawling and wriggling energy underneath the rock is at the core of its foundation. The underworld is full of activity, confusing, and regularly considered "gross."

Paying attention to DIY practices is akin to studying insects wriggling under a rock. There are no obvious starting points, no single entity is in control, and it is unclear what is going on. On first look it mostly appears to be random scurrying. Taking time and watching, however, gives more indication as to who is doing what. Ants behave in a certain way, while earthworms follow a different logic. The same applies to DIY practices: there are thousands of different types of DIY practices, and each has its own form of mindset, movement, and embodiment. The portraits drawn in these pages highlight a few species from the field of electronic and media art, and my hope is that the terms and frameworks provided will help carve out and portray a clearer picture of what goes on underneath the rock of mainline technologies and mainline art histories, so to speak. This shift in focus from monolithic "rocks" to countercurrent "bugs" has the benefit of opening up thousands of new innovative worlds. The bugs are where the action is.

But we are talking about individuals and groups of people here, not bugs. The human component cannot be emphasized enough: these are real people with warts, lumps, and their own talents and difficulties. They are real. DIY activity is a messy human activity, and highlighting it humanizes our understanding of the world and the rich subcultures within it. This is part of why DIY is wonderful—it is almost as difficult to understand as people.

The Revolutionary Potential of DIY Practices

DIY work is revolutionary. Or at least it very much tries to be. Because of its generally ad hoc approaches that lack consideration for how things are properly done, it regularly breaks disciplinary rules. This is most often a failure, but the hybrid vigor of this interbreeding of materials and ideas can also produce interesting, unicorn-like inventions. They are both messed up and brilliant. As a result, DIY approaches are useful in disrupting things—whether with technology, with organizations, or through larger culture.

In some ways we can think of DIY work as unconventional, or work with less structure. Diamonds form inside of chaos, however, and these DIY diamonds shine because they have formed out of relative entropy with few disciplinary or industry guardrails. It is a premise that even startup folks understand: if you want disruptive technologies, you need to drastically mess and shake things up.[3]

DIY-style approaches are good because they are unconventional. And they are needed given how messed up and complex the world is. The "wicked problems" that have largely stumped humanity—like climate change, social justice, homelessness, and pandemics—are too complex for a single discipline. Solving these problems requires multiple perspectives, and a useful perspective comes from disciplinary outsiders and independent polymaths mucking away on their own. DIY is not the only way—we absolutely need scientific expertise, strong regulations to protect citizens, and structure—but in the domain of innovative ideas and objects, DIY approaches produce unique results. These unique results are like raw nuggets of innovation scattered among the detritus of experimentation.

In many ways, the world demands radical interdisciplinarity—the planet has no inherent academic disciplines. Although not a panacea or cure-all, DIY practices play a core role in our innovation landscape. It can be thought of this way: every discipline used to be a DIY practice, and if we want new disciplines, we need to keep pushing ahead with DIY trajectories. The edge of a discipline is a bit like the edge of knowledge that tapers down like a landmass into an ocean of chaos. But in that ocean, way deep in a trench are things like geothermal vents that spew out toxic gas. Around those vents are radically strange ecosystems that have evolved with an aberrant logic. Giant, red-tipped tube worms, wraithlike fish, and perplexing shrimp with eyes on their backs frolic around in water hot enough to melt lead. The sphere of DIY practice is similar, and DIY culture gives birth to fantastic chimera. The projects discussed here are the offspring of electronic technology venting into the trench of art.

DIY practices like the ones outlined in this book are important because of their ubiquity, innovativeness, ability to be a voice for the marginalized, and potential to disrupt culture. DIY approaches are not creative elixirs for producing things, but rather thorny and bumpy worlds with occasional oases of inspiration. DIY is normal, awesome, desperate, brilliant, crappy, and the best thing ever. It's a fucked-up mess just like us.

The DIY Mindset: Next Steps

So, make things. Build things. Try building things that you have no qualifications for. Failure is learning. Try again. Imperfection is okay. Expose yourself to new tools, materials, places, ideas, and people. Listen and learn. Take on the humble attitude of being a nonexpert, or better yet, learn something new. Do not be afraid to humiliate yourself with technology.

Technology is only a tool, and it is yours to use. Smash and remix its pieces. Bend technology into strange and elegant things. Use DIY to extend yourself, your body, and your mind. Take it on not as a kind of libertarian cowboy, but more as a curious learner with a growth mindset, humbly and consistently forcing yourself into situations where you feel unqualified for the job but are still hacking away at it. Work on resilience and lifelong learning. Take what you've done and be proud of it, even if it's not very good.

For this attitude, the punk movement is worth taking notes from. Kim Gordon from the band Sonic Youth said, "When punk rock happened, it created an opening in the culture. . . . It made it OK to think you could play music, even though you had no musical training."[4] Kurt Cobain added, "Punk rock should mean freedom: liking and accepting anything that you like, playing whatever you want, as sloppy as you want. As long as it's good and it has passion."[5] Art comes from an informal, playful approach, so go for it. In life and all other things, it is vital to remember that passion and humanity is key.

So where do we go from here? Is the maker movement officially dead? What are the next steps?

I have no real idea what is next. It is not like DIY practice is a planned thing, anyhow. However, I have been working in the region of DIY electronic art since the early 1990s—I have seen people like Dale Dougherty from *Make* magazine come and go—and I understand parts of the terrain intimately well. And although I do not know the future, I have a number of hunches based on today's context and my knowledge of trends as a researcher in cultural uses of technology.

Nadya Tolokonnikova of Pussy Riot places an emphasis on growth in DIY culture, which is considerably more valuable to us today than anything *Make* magazine has published: "Being a punk is about constantly surprising . . . Being a punk means systematically changing the image of yourself, being elusive, sabotaging cultural and political codes . . . Punk is a method . . . [to] undermine, transform, exceed expectations. That's what punk means to me."[6] Looking forward, my advice to you is to take Tolokonnikova's attitude and apply it to the technological world. Be inspired by Survival Research Labs, Laura Kikauka, Nancy Paterson, Nam June Paik, and others. Find others like you and hang out. DIY electronic art is about constantly surprising, systematically changing your format, and sabotaging cultural and political codes. DIY electronics can be used in creative ways to undermine, transform, and exceed expectations. You have what it takes.

And this is where you come in—please let me know about it. If you are experimenting with electronic art, I invite you to literally take a photo of what you've done, print it out, scribble some notes about it, and put it in an envelope. Find a stamp and write the following on the outside:

To: Garnet Hertz
Emily Carr University of Art and Design
520 East 1st Avenue
Vancouver, BC
V5T 0H2, Canada

If you include your return address, I will try to snail mail you back something similar that I've made. For real. For fun. Or that is my hope for as long as I'm reachable at this address. I'd love to know what you're working on and learn about what you are going through, and I will consider your contribution as an unofficial addition to this book. We have surveyed only a minor slice of this exciting terrain, and I invite you—wherever you are and whatever your qualifications—to help organize, write, and build the future of electronic art, punk technologies, critical making, and the DIY mindset together with me. I have no idea what this will look like, but this is the amazingly fun part. Let's try to connect at some point and figure it out.

So go boldly and make things with passion. This includes reinventing the technologies around us that sculpt our everyday phones, videos, music, appliances, communication technologies, or whatever. Reinvent it all if you like. Nobody is qualified to do something that hasn't been done anyhow. I'll help you if I can. The heart of the matter is that humanity is capable of so much more than only making money or engineering drabness to be bought and sold at places like Walmart. Those places don't reflect the best stuff in life. Your humble DIY project is honestly better.

Do it yourself, do it with your friends and loved ones, and do it to make things better for more than just yourself. Make messed up and weird shit with raw humanity. Hack tech hard. Give a damn about others and be kind. Put your heart into it. Listen. Learn the humiliation of learning something new, and find the beauty within mistakes. Try not to be afraid as you figure it out as a lone oddball: you are beautiful, and you can do it yourself.

Notes

Beginnings

1. Michel Foucault, "Heterotopias (Of Other Spaces, 1967)," http://individual.utoronto.ca/bmclean/hermeneutics/foucault_suppl/heterotopias.html. And Michel Foucault, *The Order of Things: An Archaeology of the Human Sciences* (New York: Vintage Books, 1973).

2. Usenet was a decentralized computer network for people to discuss specific topics and to share files in newsgroups.

3. Mark Pauline, "Beyond the Realm of Humans: A Discussion with Mark Pauline of Survival Research Laboratories," interview by Garnet Hertz, conceptlab.com, 1995, https://www.conceptlab.com/interviews/pauline.html.

4. Billy Kluver, Julie Martin, and Barbara Rose, eds., *Pavilion: Experiments in Art and Technology* (Dutton, 1972).

5. Heterotopia is a term used by Foucault to describe the antithesis of utopia. Heterotopias are typically troubling, exceptional, contrasting, or transformative environments. They are worlds within worlds, mirroring but also inverting normal society. Foucault provides several examples including boats, graveyards, bars, brothels, jails, and fairs. Foucault, *The Order of Things*.

6. Frank Popper, *Art of the Electronic Age* (New York: Harry N Abrams Inc, 1993).

7. Ken Goldberg, ed., *The Robot in the Garden: Telerobotics and Telepistemology in the Age of the Internet*, Leonardo (Cambridge, MA: MIT Press, 2000).

8. *The Telegarden* launched in June 1995 at the University of Southern California where Goldberg was teaching, and in September 1996 the project moved to the Ars Electronica Center and operated until August 2004. "The Telegarden," (website), accessed August 16, 2021, https://goldberg.berkeley.edu/garden/Ars/.

9. Peter H. Kahn Jr., *Technological Nature: Adaptation and the Future of Human Life* (Cambridge, MA: MIT Press, 2011), 151.

10. "INTERFACE—Webmachine Project Archive [Formerly Known as Doppelganger]," conceptlab.com, Garnet Hertz, accessed August 16, 2021, https://conceptlab.com/interface/. Also see Eduardo Kac, "Towards a Chronology of Robotic Art," *Convergence* 7, no. 1 (March 2001): 87–111, https://doi.org/10.1177/135485650100700109.

11. Rafael Lozano-Hemmer, "Perverting Technological Correctness," *Leonardo* 29, no. 1 (1996): 5, https://doi.org/10.2307/1576269.

12. Anthony Dunne, *Hertzian Tales: Electronic Products, Aesthetic Experience and Critical Design* (RCA CRD Research Publications, 1999).

13. Popper, *Art of the Electronic Age*.

14. See documentation of the Center for Metahuman Exploration's *Petting Zoo* project that was featured at ISEA97 in Chicago. *Petting Zoo*, The Center for Metahuman Exploration, accessed August 16, 2021, https://metahuman.bitbucket.io/web/pzoo1.html.

15. For additional information on the ISEA97 event, see website for conference proceedings. "ISEA97 Chicago—Conference Proceedings," ISEA Symposium Archives, accessed August 16, 2021, http://www.isea-archives.org/docs/1997/proceedings/ISEA97_proceedings.pdf.

16. Eduardo Kac, *Art at the Biobotic Frontier* (Chicago: The School of the Art Institute of Chicago, 1997), accessed August 16, 2021, http://ekac.org/apositive.html.

17. For documentation of this project—*La Cour des Miracles*—see *La Cour des Miracles*, Bill Vorn, accessed August 16, 2021, https://billvorn.concordia.ca/robography/Cour.html. Or Louis-Philippe Demers, *La Cour des Miracles*, 2015, video, 7:02, https://vimeo.com/127682457.

18. Davies presented the project "*Osmose*" during the panel "Beyond Shelter: The Future of Architecture" at ISEA97. "Paper: Char Davies—Osmose," ISEA Symposium Archives, accessed August 16, 2021, http://www.isea-archives.org/isea96-john-law-drawing-on-the-brain-an-art-science-collaboration-2-2-2-2-2-2-2-2-2-2-2-2-2-2-2-2-2-2-2-2-2-2-2-2-2-2-2-2-2-2-2-2-2-2-2-2-2-2-2-2-2-2-2-2-2-2-2-2-2-2-2-2-2-2-2-2-2-2-2-2-2-2-2-2-98/.

19. Silvia Lindtner, Garnet D. Hertz, and Paul Dourish, "Emerging Sites of HCI Innovation: Hackerspaces, Hardware Startups & Incubators," in *CHI '14: Proceedings of the SIGCHI Conference on Human Factors in Computing Systems* (New York, NY: Association for Computing Machinery, 2014), 439–448, https://doi.org/10.1145/2556288.2557132.

20. Tim O'Reilly, "Why We Spun out Maker Media," Radar, O'Reilly, accessed August 16, 2021, http://radar.oreilly.com/2013/01/why-we-spun-out-maker-media.html.

21. Zoe Sofia, "Contested Zones: Futurity and Technological Art," *Leonardo* 29, no. 1 (1996), 504.

22. Lindtner, Hertz, and Dourish, "Emerging Sites of HCI Innovation, 439–448, https://doi.org/10.1145/2556288.2557132.

23. Stephen Wilson, *Information Arts: Intersections of Art, Science, and Technology* Leonardo (Cambridge, MA: MIT Press, 2001).

24. The use of straightforward language in academic writing here is primarily motivated by a desire for legibility by a wider audience. The decision to stay away from the well-worn names of critical theory when writing about art—whether invoking Foucault, Derrida, Barthes, Adorno, Horkheimer, Jung, or Freud—is motivated by a desire to stretch the discipline more in the direction of a publicly accessible design theory and art history. Instead of using complex academic language, this project works to pull together widely disparate and complex academic fields in a clear, cohesive, and memorable way. Or at least, that's the core idea.

0.1 A History of Electronic Art in the Twentieth Century

1. According to the Museum of Modern Art, *Objet Trouvé* is defined as a "term applied in the 20th century to existing objects, manufactured or of natural origin, used in, or as, works of art. With the exception of the Ready-made, in which a manufactured object is generally presented on its own without mediation, the *objet trouvé* is most often used as raw material in an Assemblage, with juxtaposition as a guiding principle." Matthew Gale, "Objet trouvé," Art Terms, The Museum of Modern Art, 2009, archived June 20, 2010, https://web.archive.org/web/20100620150735/http://moma.org/collection/theme.php?theme_id=10135.

2. For example, see Kurt Schwitters, "Merz Picture 32 A. The Cherry Picture (Merzbild 32 A. Das Kirschbild). 1921," Art and artists, The Museum of Modern Art, https://www.moma.org/collection/works/33356.

3. Violetta Farina and Johanna Kreiner, *20th Century Art* (Florence, IT: SCALA, 2009), 309.

4. In a contemporary American context, an appropriate object might be a plastic lawn chair from Walmart.

5. Marcel Duchamp, "Joan Bakewell in Conversation with Marcel Duchamp," interview by Joan Bakewell, *Late Night Line-Up*, BBC Arts, 1968, accessed August 17, 2021, https://www.bbc.co.uk/programmes/p04826th.

6. Robert Fulford, "Robert Fulford: If Anything's Art, Art's Nothing," *National Post*, May 3, 2015, https://nationalpost.com/opinion/robert-fulford-if-anythings-art-arts-nothing.

7. Diane Waldman, *Collage, Assemblage, and the Found Object* (New York: Harry N. Abrams Inc, 1992), 17.

8. This piece has a body of recent scholarship that indicates that Duchamp had considerable external help in its development, with the Baroness Elsa recognized as the likely originator of the idea of using a urinal as a sculpture. Irene Gammel, *Baroness Elsa: Gender, Dada, and Everyday Modernity: A Cultural Biography* (Cambridge, MA: MIT Press, 2002), 224–25.

9. Conceptual Art can be described as art practice that gives priority to ideas over objects, to concepts over the physical form of the work, and to process over the resulting end product. Philip Hensher, "The Loo That Shook the World: Duchamp, Man Ray, Picabi," *The Independent*, February 20, 2008, 2–5.

10. Marcel Duchamp, "Rotary Demisphere (Precision Optics). Paris, 1925," Art and artists, The Museum of Modern Art, accessed August 17, 2021, https://www.moma.org/collection/works/81432.

11. Spencer Tucker, *The Great War, 1914–18* (Bloomington: Indiana University Press, 1998), 11.

12. Between nine and fifteen million people were killed in the First World War. Matthew White, "Source List and Detailed Death Tolls for the Primary Megadeaths of the Twentieth Century," last updated February 2011, http://necrometrics.com/20c5m.htm#WW1.

13. Marcel Janco, "Dada at Two Speeds," in *Dadas on Art: Tzara, Arp, Duchamp and Others*, ed. Lucy R. Lippard (Courier Corporation, 2007), 36.

14. Leah Dickerman, Brigid Doherty, Centre Georges Pompidou, U.S. National Gallery of Art, Museum of Modern Art, *Dada: Zurich, Berlin, Hannover, Cologne, New York, Paris*. Distributed Art Pub Incorporated, 2005.

15. Peter Wiebel, "It Is Forbidden Not to Touch: Some Remarks on the (Forgotten Parts of the) History of Interactivity and Virtuality," in *MediaArtHistories*, Leonardo, ed. Oliver Grau (Cambridge, MA: MIT Press, 2007), 26.

16. Henry Lie, "Replicas of László Moholy-Nagy's Light Prop: Busch-Reisinger Museum and Harvard University Art Museums," Tate Papers, Tate, 2007, https://www.tate.org.uk/research/publications/tate-papers/08/replicas-of-laszlo-moholy-nagys-light-prop-busch-reisinger-museum-and-harvard-university-art-museums.

17. Bruno Munari and Umberto Eco, quoted in Frank Popper, *Origins and Development of Kinetic Art* (New York Graphic Society, 1968), 104.

18. Norbert Wiener, *Cybernetics or Control and Communication in the Animal and the Machine, Reissue of the 1961 Second Edition* (1961; repr., Cambridge, MA: MIT Press, 2019).

19. C. E. Shannon, "A Mathematical Theory of Communication," *The Bell System Technical Journal* 27, no. 3 (July 1948): 379–423, https://doi.org/10.1002/j.1538-7305.1948.tb01338.x.

20. Edward A Shanken, "Art in the Information Age: Cybernetics, Software, Telematics and the Conceptual Contributions of Art and Technology to Art History and Aesthetic Theory" (PhD diss., Duke University, 2001), 23, https://artexetra.files.wordpress.com/2010/04/shanken_art_info_age_diss_2001.pdf.

21. Norbert Wiener, *The Human Use of Human Beings: Cybernetics and Society,* rev. ed. (New York, NY: Da Capo Press, 1988).

22. Walter J. Ong, *Orality and Literacy: The Technologizing of the Word* (Psychology Press, 2002), 2.

23. The exhibition ran from November 27, 1968, to February 9, 1969.

24. The exhibition included a large number of artists including Malevich, Tatlin, Duchamp, Klee, Gabo, Moholy-Nagy, Picabia, Calder, Tinguely, Oldenburg, Takis, and Tinguely. K. Pontus Hultén, *The Machine as Seen at the End of the Mechanical Age* (New York, NY: Museum of Modern Art, 1968), exhibition catalog distributed by New York Graphic Society Ltd.

25. Museum of Modern Art, "The Machine as Seen at the End of the Mechanical Age," press release no. 123, November 27, 1968, https://www.moma.org/documents/moma_press-release_326596.pdf. More information about the exhibition can be found at "The Machine as Seen at the End of the Mechanical Age, November 27, 1968–February 9, 1969, MoMA," accessed August 17, 2021, https://www.moma.org/calendar/exhibitions/2776.

26. Jasia Reichardt, ed., *Cybernetic Serendipity: The Computer and the Arts* (New York, NY: Studio International, July 1968), https://monoskop.org/images/2/25/Reichardt_Jasia_ed_Cybernetic_Serendipidity_The_Computer_and_the_Arts.pdf.

27. The term "electronic art" also had emerged before the MoMA "Machine" and ICA "Cybernetic" exhibitions. Nam June Paik, for example, had used the term to describe his work since at least 1965. Paik did three solo exhibitions between 1965 and 1971 at Galeria Bonino in New York City titled "Electronic Art" (1965), "Electronic Art II" (1968), and "Electronic Art III" (1971)—two of which predate the MoMA "Machine" exhibition.

28. Fred Turner, *From Counterculture to Cyberculture: Stewart Brand, the Whole Earth Network, and the Rise of Digital Utopianism* (Chicago: University of Chicago Press, 2006), 38.

29. Richard Rinehart and Jon Ippolito, "Death by Technology," in *Re-Collection: Art, New Media, and Social Memory,* Leonardo (Cambridge, MA: MIT Press, 2014), 31–33.

30. Shanken, "Art in the Information Age," 11–12.

31. Shanken, 147.

32. Richard Rinehart and Jon Ippolito, *Re-Collection: Art, New Media, and Social Memory*, Leonardo (Cambridge, MA: MIT Press, 2014), 32.

33. Craig Harris, ed., "The Xerox Palo Alto Research Center Artist-in-Residence Program Landscape," in *Art and Innovation: The Xerox PARC Artist-in-Residence Program* (Cambridge, MA: MIT Press, 1999), 9, https://direct.mit.edu/books/book/2566/Art-and-InnovationThe-Xerox-PARC-Artist-in.

34. Gloria Bryant, "The Switched-on Theater," in *Bell Telephone Laboratories Reporter* 15, No. 6. (November–December 1966): 12–17.

35. Clive Barnes, "Dance or Something at the Armory; New Series Presents Rauschenberg Work Audience Endures a Depressing Spectacle," *New York Times*, October 15, 1966, Archives, https://www.nytimes.com/1966/10/15/archives/dance-or-something-at-the-armory-new-series-presents-rauschenberg.html.

36. The not-for-profit organization had Robert Rauschenberg as chair, Billy Klüver as president, Robert Whitman as treasurer, and Fred Waldhauer as secretary. Experiments in Art and Technology, "Experiments in Art and Technology: A Brief History and Summary of Major Projects," March 1, 1998, 1, Cinémathèque québécoise, Montréal; and Kristine Stiles and Peter Selz, *Theories and Documents of Contemporary Art: A Sourcebook of Artists' Writings*, 2nd ed. (Berkeley, CA: University of California Press, 2012), 453.

37. Experiments in Art and Technology, "Experiments in Art and Technology: An Introduction and Report on Current Activities," May 4, 1971, 2, Cinémathèque québécoise, Montréal. Thanks to Cinémathèque québécoise for generously providing access to their archives.

38. Marshall McLuhan, *Understanding Media: The Extensions of Man*. (New York: McGraw-Hill, 1964) xi.

39. Experiments in Art and Technology, "Experiments in Art and Technology: A Brief History and Summary of Major Projects 1966–1998," March 1st, 1998, Preface, Cinémathèque québécoise, Montréal.

40. Billy Klüver, letter to Boyd Compton of the Rockefeller Foundation, March 24, 1966, Collection of the Cinémathèque québécoise, Montréal.

41. Experiments in Art and Technology, "Experiments in Art and Technology: A Brief History and Summary of Major Projects 1966–1998," March 1st, 1998, 2, Cinémathèque québécoise, Montréal.

42. Frances Dyson, *The Tone of Our Times: Sound, Sense, Economy, and Ecology*, Leonardo (Cambridge, MA: MIT Press, 2014), 143.

43. Jasia Reichardt, "Art at Large," *New Scientist* 54, no. 790 (April 6, 1972): 37.

44. Barnes, "Dance or Something at the Armory."

45. Marshall McLuhan, *The Gutenberg Galaxy: The Making of Typographic Man* (University of Toronto Press, 1962), 31.

46. McLuhan, *Understanding Media,* 3.

47. McLuhan, xi.

48. Arjen Mulder and Maaike Post, *Book for the Electronic Arts* (V2_ publishing, 2000), 81–87.

49. "InterAccess | Non-Profit Gallery, Educational Facility, Production Studio, and Festival," accessed August 17, 2021, https://interaccess.org/.

50. Raymond Bellour and Bill Viola, "An Interview with Bill Viola," *October* 34 (1985): 91–119, https://doi.org/10.2307/778491. See Alexander R. Galloway, *Protocol: How Control Exists after Decentralization,* Leonardo (Cambridge, MA: MIT Press, 2004), 211.

51. Shanken, "Art in the Information Age," 84.

52. Robert Smithson putting mirrors in the desert was sort of a do-it-yourself form of land art. Johannes Stückelberger, "Mirror Reflections: Robert Smithson's Dialectical Concept of Space," *RACAR: Revue d'art Canadienne* 31, no. 1–2 (2006): 90, https://doi.org/10.7202/1069626ar.

53. Klüver writes: "E.A.T. has essentially achieved its goal of fostering an environment in which the artist is able to incorporate new technology in his work. Thirty years ago the separation of art and technology seemed insurmountable. Today artists find themselves at ease with the incorporation of technology in their works. Of course, E.A.T. cannot take full credit for this, and some social problems remain to be dealt with in the art world. One of the problems is with the museums, which are unprepared, even today, to receive works involving technology. Curators tend to file away 'difficult' works. If a visitor says, 'I saw it, but it didn't work,' the situation is not acceptable. A few museums, like the Centre George Pompidou, have attached engineers which service such works, and many of the younger curators have gotten the point." Billy Klüver, undated letter, estimated after 1996 and before 1999, E.A.T. Today, Collection of the Cinémathèque québécoise, Montréal. Klüver was most likely referring to Alain Peron, who worked at the Centre Pompidou between 1982 and 2011 as "technicien audiovisuel."

54. Granted, electricity and electromechanics are approximately a billion years older than electronics.

55. Canada Council for the Arts, "Fields of Practice," accessed July 27, 2022, https://canadacouncil.ca/funding/grants/guide/apply-to-programs/fields-of-practice.

56. For example: "the medium is the message . . . the personal and social consequences of any medium—that is, of any extension of ourselves—result from the new scale that is introduced into our affairs by each extension of ourselves, or by any new technology." Eric Mcluhan and Frank Zingrone, *Essential McLuhan* (Concord, Ont: House of Anansi Press, 1995), 15.

57. For an overview of digital art, see Christiane Paul, *Digital Art* (London: Thames & Hudson, 2006), 7–8.

58. Beryl Graham and Sarah Cook, *Rethinking Curating: Art after New Media*, Leonardo (Cambridge, MA: MIT Press, 2010), 10.

59. For an overview of Holopoetry, see Simone Osthoff, "12: From Mail Art to Telepresence: Communication at a Distance in the Works of Paulo Bruscky and Eduardo Kac," in *At a Distance: Precursors to Art and Activism on the Internet*, eds. Annmarie Chandler and Norie Neumark (Cambridge, MA: MIT Press, 2005), 269–270, https://doi.org/10.7551/mitpress/1467.001.0001.

60. Peter H Rossi, Bruce J Biddle, and Neal Balanoff, *The New Media and Education; Their Impact on Society* (Chicago: Aldine Pub. Co., 1966).

61. Rafael Lozano-Hemmer, "Perverting Technological Correctness," *Leonardo* 29, no. 1 (1996): 5–15.

62. Lozano-Hemmer, "Perverting Technological Correctness," 6.

63. Silvia Lindtner, Shaowen Bardzell, and Jeffrey Bardzell, "Reconstituting the Utopian Vision of Making: HCI After Technosolutionism," *Proceedings of the 2016 CHI Conference on Human Factors in Computing Systems* (May 2016): 1390–1402, https://doi.org/10.1145/2858036.2858506.

64. Author's private email correspondence with Lorne Falk, March 7th, 2018.

65. Mulder and Post, *Book for the Electronic Arts*, 85.

66. David Richard Rokeby, Su Ditta, and Sara Diamond, *David Rokeby: Oakville Galleries, 25 June to 17 October 2004* (Oakville Galleries, 2004), 73–74.

67. Shanken, "Art in the Information Age," 433–438.

68. Online social networking and user-generated content services were rapidly expanding during this period. Friendster was released in 2002, MySpace in 2003, del.icio.us in 2003, Orkut in 2004, and Facebook in 2004. Soon after this, YouTube was released in 2005, and Twitter launched in 2006.

69. Peter Lunenfeld, ed., *The Digital Dialectic: New Essays on New Media*, Leonardo (Cambridge, MA: MIT Press, 1999), xvi.

70. Anthony Dunne, *Hertzian Tales: Electronic Products, Aesthetic Experience, and Critical Design* (Cambridge, MA: MIT Press, 2006), 3.

71. This is somewhat understandable considering that Dunne's dismissal was written at the height of dot-com excitement in 1999.

72. See the following for a description of boundary negotiating artifacts: Charlotte P. Lee, "Between Chaos and Routine: Boundary Negotiating Artifacts in Collaboration," in *ECSCW 2005*, ed. Hans Gellersen et al. (Dordrecht: Springer Netherlands, 2005), 387–406, https://doi.org/10.1007/1-4020-4023-7_20. This term is an offshoot of "boundary object" popularized by Star and Bowker. See Susan Leigh Star and James R. Griesemer, "Institutional Ecology, 'Translations' and Boundary Objects: Amateurs and Professionals in Berkeley's Museum of Vertebrate Zoology, 1907–39," *Social Studies of Science* 19, no. 3 (August 1, 1989): 387–420, https://doi.org/10.1177/030631289019003001.

73. Machiko Kusahara, "Device Art: A New Form of Media Art from a Japanese Perspective," *Intelligent Agent* 06, no. 2 (2006), https://www.semanticscholar.org/paper/Device-Art-%3A-A-New-Form-of-Media-Art-from-a-Kusahara/e4566934d6675a3d5649d3da4acd9c20fa009bb9.

0.2 A Definition of DIY

1. Florian Cramer, "Does DIY Mean Anything?—A DIY Attempt" (essay, originally commissioned for Anrikningsverket Journal #1 by Norbergfestival, July 2019), http://cramer.pleintekst.nl/essays/does_diy_mean_anything/.

2. Julia Lupton, "D.I.Y. Theory," in *DIY: Design It Yourself: A Design Handbook,* ed. Ellen Lupton (Princeton, NJ: Princeton Architectural Press, 2006). Chapter available online at http://www.designwritingresearch.org/mica_diy/master/Theory.pdf

3. C. Simionato, "Europe: Leading DIY Retailers 2019, by Revenue," Statista, May 11, 2021, https://www.statista.com/statistics/600120/diy-stores-net-turnover-in-europe/.

4. Chuck Welch, ed., *Eternal Network: A Mail Art Anthology* (Calgary: University of Calgary Press, 1994).

5. Paul Dourish, *Where the Action Is: The Foundations of Embodied Interaction* (Cambridge, MA: MIT Press, 2001), 126.

6. Matthew B. Crawford, *Shop Class as Soulcraft: An Inquiry into the Value of Work* (New York: Penguin Books, 2009), 27.

7. Richard Sennett, ed., *The Culture of the New Capitalism* (New Haven, CT: Yale University Press, 2007), 194.

8. Ann Cvetkovich, *Depression: A Public Feeling* (Durham, NC: Duke University Press, 2012), 168.

9. William Gibson, *Neuromancer* (New York: Ace Books, 1984).

10. The term "meatspace"—according to Merriam-Webster—originated online in Usenet discussion groups in the early 1990s and spread to the popular press by 1995. "What Is 'Meatspace'?," Words We're Watching, Merriam-Webster, accessed August 18, 2021, https://www.merriam-webster.com/words-at-play/what-is-meatspace.

11. John Jordan, "The Art of Necessity: The Subversive Imagination of Anti-road Protest and Reclaim the Streets," in *Cultural Resistance Reader*, ed. Stephen Duncombe (London: Verso, 2002), 347–357.

12. Amy Spencer, *DIY: The Rise of Lo-Fi Culture*, 2nd ed. (London: Marion Boyars, 2008), 11.

13. Claude Lévi-Strauss, *The Savage Mind* (Chicago, IL: The University of Chicago Press, 1966), 17–18.

14. Spencer also sees DIY processes as "not trying to seek out new technology." This is partially incorrect, however, since the tools or "it" of DIY can be sought out. For example, someone wanting to make custom T-shirts may teach themselves silkscreening. New technologies or tools can be chosen, but DIY practitioners tend to remain outsiders to the standard way of doing things. By this I mean that they implement nonstandard methodologies, or are actively ignorant of the official way those methodologies are supposed to be used. Several case studies discussed in the following chapters include artists that intentionally tackle the medium of working with electronics, and most of those artists work outside of the professional sphere of being an electrician or an electrical engineer.

15. Vanessa Beatty, *20 Fun and Creative Crafts with Plastic Soda Bottles*, March 4, 2014, https://www.diyncrafts/com/4870/.

16. See chapter 4.3 of this book.

17. Walter Benjamin, *The Work of Art in the Age of Its Technological Reproducibility, and Other Writings on Media* (Cambridge, MA: Belknap Press of Harvard University Press, 2008), 86.

18. Graham Harman, *Tool-Being: Heidegger and the Metaphysics of Objects* (Chicago, IL: Open Court, 2002).

19. Ian Bogost, "What Is Object-Oriented Ontology? A Definition for Ordinary Folk," *Writing* (blog), December 8, 2009, http://bogost.com/writing/blog/what_is_objectoriented_ontolog/.

20. Jane Bennett, *Vibrant Matter: A Political Ecology of Things* (Durham, NC: Duke University Press, 2010), viii.

21. Emit Snake-Beings, "The DiY ['Do It Yourself'] Ethos: A Participatory Culture of Material Engagement" (PhD diss., Hamilton, New Zealand, University of Waikato, 2016), https://hdl.handle.net/10289/9973.

22. Rachel Maines, *Hedonizing Technologies: Paths to Pleasure in Hobbies and Leisure* (Baltimore, MD: Johns Hopkins University Press, 2009).

23. Lisa Gitelman, *Paper Knowledge: Toward a Media History of Documents* (Durham, NC: Duke University Press, 2014), 140–101, https://doi.org/10.1515/9780822376767; and Karen Sánchez-Eppler, *Dependent States: The Child's Part in Nineteenth-Century American Culture* (Chicago, IL: University of Chicago Press, 2005), 153.

24. Cramer, "Does DIY Mean Anything?"

25. Arthur G. Bedeian and Daniel A. Wren, "Most Influential Management Books of the 20th Century," *Organizational Dynamics* 29, no. 3 (2001): 222.

26. Frederick Winslow Taylor, *The Principles of Scientific Management* (New York: Harper & Bros., 1919), 83. Originally self-published in 1911.

27. Taylor claims that workers using his method receive between 30 and 100 percent more pay than workers of a similar caliber. Frederick Winslow Taylor, *The Principles of Scientific Management* (New York: Harper & Bros., 1919), 28. Originally self-published in 1911.

28. Taylor, *Principles of Scientific Management*, 83.

29. Tony Swan, "Ford's Assembly Line Turns 100: How It Really Put the World on Wheels," *Car and Driver*, April 30, 2013, https://www.caranddriver.com/features/a15115930/fords-assembly-line-turns-100-how-it-really-put-the-world-on-wheels-feature.

30. Georges Friedmann, *The Anatomy of Work: Labor, Leisure, and the Implications of Automation*, trans. Wyatt Rawson (New Brunswick, NJ: Transaction Publishers, 1992), xx.

31. Taylor, *Principles of Scientific Management*, 5.

32. Anthony Dunne, *Hertzian Tales: Electronic Products, Aesthetic Experience, and Critical Design* (Cambridge, MA: MIT Press, 2006), 20.

33. Peter Dormer, *The Meanings of Modern Design: Towards the Twenty-First Century* (New York : Thames and Hudson, 1990), 124.

34. Dunne, *Hertzian Tales*, 137–142.

35. Dunne, 145.

36. Dunne, 35.

37. As arguably the most industrialized country in the world in the 1800s, "Britain was also the first to attempt to counter the malevolent aspects of mass production—the cost in human creativity and quality of life." Wendy Kaplan, *The Arts & Crafts Movement in Europe & America: Design for the Modern World* (New York, NY: Thames & Hudson in association with the Los Angeles County Museum of Art, 2004), 11.

38. Sam Jacob, "What Makes a Maker? Hint: It's Not Just About 'Making,'" *Metropolis* (blog), March 23, 2015, https://www.metropolismag.com/design/arts-culture/what-makes-a-maker/.

39. Jacob, "What Makes a Maker?"

40. Dunne, *Hertzian Tales*; Anthony Dunne and Fiona Raby, *Speculative Everything: Design, Fiction, and Social Dreaming* (Cambridge, MA: MIT Press, 2013).

Theme 1

1. Garnet Hertz and Jussi Parikka, "Zombie Media: Circuit Bending Media Archaeology into an Art Method," *Leonardo* 45, no. 5 (October 2012): 424–30, https://doi.org/10.1162/LEON_a_00438.

2. Vladimir Arkhipov and Susan Glasser, *Home-Made: Contemporary Russian Folk Artifacts*, ed. Damon Murray and Stephen Sorrell (London: FUEL Publishing, 2006).

3. Pénélope de Bozzi and Ernesto Oroza, *Objets Réinventés: La Création Populaire à Cuba* (Paris: Alternatives, 2002).

4. For an example of an online resource for farmers to share their own DIY methods and designs, see "Farm Hack," accessed August 26, 2021, https://farmhack.org/. The group describes itself as "a community for those who embrace the long-standing farm traditions of tinkering, inventing, fabricating, tweaking, and improving things that break."

5. Charles Jencks and Nathan Silver, *Adhocism: The Case for Improvision* (London: Socher & Warburgh, 1972).

6. Deepa Butoliya, "Critical Jugaad: A Postnormal Design Framework for Designerly Marginal Practices" (PhD diss., Carnegie Mellon University, 2018), 55.

7. Richard Rinehart and Jon Ippolito, eds., *Re-Collection: Art, New Media, and Social Memory*, Leonardo (Cambridge, MA: MIT Press, 2014), 31–33.

8. Richard Grusin, John G. Hanhardt, Richard Shiff, and Jim Campbell, *Jim Campbell—Material Light* (New York: Bryce Wolkowitz Gallery, 2010).

9. Thomas Thwaites, *The Toaster Project: Or a Heroic Attempt to Build a Simple Electric Appliance from Scratch* (New York: Princeton Architectural Press, 2011), 169–170.

10. Ezio Manzini, "Design Culture and Dialogic Design," *Design Issues* 32, no. 1 (January 2016): 58, https://doi.org/10.1162/DESI_a_00364.

1.1 Frugality and the *Demanufacturing Machine*

1. The *Demanufacturing Machine* was first shown on February 25, 1979. The initial *Mad Max* movie was released in Australia on April 12, 1979, and was released in the USA the following year on March 21, 1980.

2. Andrew Meacham, "James R. Carlson, Who Founded the Eckerd College Theater Program, Dies at 94," *Tampa Bay Times*, February 1, 2012, https://www.tampabay.com/news/obituaries/james-r-carlson-who-founded-the-eckerd-college-theater-program-dies-at-94/1213478/.

3. Falstaff Brewing was the third largest brewer in the United States by the 1960s, but an antitrust case brought by the state of Rhode Island after a merger with another company in 1965 set the brewery on a downward trajectory that they never recovered from. A series of other mergers in the American brewing industry in the 1970s essentially put Falstaff out of business, and the brewery in San Francisco was shuttered in 1978 after opening only three years before.

4. Sarah Minter, *Nadie es inocente* (1986), film, 57:00, from Sonido Pappos, "Nadie es Inocente M.P. By Sarah Minter 1/6," YouTube video, 9:48, April 30, 2010, https://www.youtube.com/watch?v=_zl019-59U8.

5. Genuine Survival Research Labs, "Survival Research Labs at LA MOCA Sept 30 2011," YouTube video, 14:28, October 17, 2011, https://www.youtube.com/watch?v=vcdE2Ai-TPc.

6. Mark Pauline, "Beyond the Realm of Humans: A Discussion with Mark Pauline of Survival Research Laboratories," interview by Garnet Hertz, conceptlab.com, 1995, https://www.conceptlab.com/interviews/pauline.html.

7. Pauline, "Beyond the Realm of Humans."

8. V. Vale, "Consumate SRL Show in LA," 2005, https://srl.org/shows/la2005/vale.html.

9. Richard Barbrook and Andy Cameron, "The Californian Ideology," *Science as Culture* 6, no. 1 (January 1996): 44–72, https://doi.org/10.1080/09505439609526455.

10. Frank Popper, *Origins and Development of Kinetic Art*, trans. Stephen Bann (London: Studio Vista, 1968), 131.

11. Bruce Sterling, "Is Phoenix Burning," *Wired*, July 1, 1996, https://www.wired.com/1996/07/srl/.

12. Hiroo Iwata, "What Is Device Art?," University of Tsukuba, archived February 8, 2012, https://web.archive.org/web/20120208121046/http://www.deviceart.org/.

13. Garnet Hertz and Jussi Parikka, "Zombie Media: Circuit Bending Media Archaeology into an Art Method," *Leonardo* 45, no. 5 (October 2012): 424–430, https://doi.org/10.1162/LEON_a_00438.

14. Hertz and Parikka, "Zombie Media," 429.

15. For more information on Gartner Group's Hype Cycle theory, see Mark Raskino and Jackie Fenn, "Understanding Gartner's Hype Cycles," *Gartner*, July 10, 2009, https://www.gartner.com/en/documents/1069314/understanding-gartner-s-hype-cycles-2009.

16. Claude Lévi-Strauss *The Savage Mind*, The Nature of Human Society Series (Chicago: Univ. of Chicago Press, 2000), 17–18.

17. Horning states that hipsters are an "embodiment of postmodernism as a spent force, revealing what happens when pastiche and irony exhaust themselves as aesthetics." Rob Horning, "The Death of the Hipster," PopMatters, April 13, 2009, https://www.popmatters.com/the-death-of-the-hipster-panel-2496026662.html.

18. See "The Women of SRL," Tentacle Session #35, July 21, 2002, http://sessions.laughingsquid.org/past/SRLWomen.html, and "Laughing Squid's Tentacle Sessions #35: Women in SRL," accessed August 19, 2021, https://srl.org/shows/events/tentacle/women.html for details on female involvement in Survival Research Laboratories.

19. Eric Paulos, "SRL Information Update," July 15, 1994, https://www.srl.org/information/srl_info.html.

20. Nicholas Rombes, *A Cultural Dictionary of Punk: 1974–1982* (New York: Continuum, 2009), 201–2.

21. Paulos, "SRL Information Update."

22. See https://srl.org/shows/austin/netcredits.html for more information.

23. See https://srl.org/lab.html for more information.

24. The array of non-white/cis/male/straight punk scholarship is broad and expanding. See work by Mike Dines, Alastair Gordon, Paula Guerra, Russ Bestley, José Esteban Muñoz, Tom Jennings, Earl Miller, Mimi Thi Nguyen, Larry-Bob Roberts, Bruce La Bruce, C. Ondine Chavoya, David Evans Frantz, and the KISMIF ("'Keep It Simple, Make It Fast") Conference.

25. Michel de Certeau, *The Practice of Everyday Life*, trans. Steven Rendall, (1984; repr., Berkeley: Univ. of California Press, 2002), xiv.

1.2 Frugality and *Telephonic Arm-Wrestling*

1. Charles Jencks and Nathan Silver,"Foreword to the MIT Press Edition: The Style of Eureka," in *Adhocism: The Case for Improvisation* (Cambridge, MA: MIT Press, 2013), vii–xxii.

2. Amit S Rai, "The Affect of Jugaad: Frugal Innovation and Postcolonial Practice in India's Mobile Phone Ecology," *Environment and Planning D: Society and Space* 33, no. 6 (December 2015): 985–1002, https://doi.org/10.1177/0263775815598155.

3. Kavita Philip, Lilly Irani, and Paul Dourish, "Postcolonial Computing: A Tactical Survey," *Science, Technology, & Human Values* 37, no. 1 (January 2012): 3–29, https://doi.org/10.1177/0162243910389594.

4. Philip, Irani, and Dourish, "Postcolonial Computing," 3–29.

5. Nimmi Rangaswamy and Nithya Sambasivan, "Cutting Chai, Jugaad, and Here Pheri: Towards UbiComp for a Global Community," *Personal and Ubiquitous Computing* 15, no. 6 (August 2011): 553–64, https://doi.org/10.1007/s00779-010-0349-x.

6. Beatrice Jauregui, "Provisional Agency in India: *Jugaad* and Legitimation of Corruption," *American Ethnologist* 41, no. 1 (February 2014): 76–91, https://doi.org/10.1111/amet.12061.

7. Jauregui, "Provisional Agency in India."

8. Deepa Butoliya, "Critical Jugaad: A Postnormal Design Framework for Designerly Marginal Practices" (PhD diss., Carnegie Mellon University, 2018), 55.

9. Butoliya, 71–75.

10. See https://gasworks.org.uk/residencies/yashas-shetty/ for more information.

11. Exhibition Catalog, *Gambiólogos: A Gambiarra Nos Tempos Do Digital* (Kludging in a Digital Era, Collective Exhibition, 2010), https://issuu.com/gambiologia/docs/catalogo_gambiologos_web/49.

12. Navi Radjou, Jaideep Prabhu, and Simone Ahuja, *Jugaad Innovation: A Frugal and Flexible Approach to Innovation for the 21st Century* (San Francisco: Random House India, 2012), 5.

13. Pénélope de Bozzi and Ernesto Oroza, *Objets Réinventés: La Création Populaire à Cuba* (Paris: Alternatives, 2002).

14. Norman White and Doug Back, "Telephonic Arm-Wrestling," Works, V2_Lab for the Unstable Media, accessed August 19, 2021, https://v2.nl/archive/works/telephonic-arm-wrestling.

15. V2_, "Tele-Armwrestling by Norman White (2011)," video, 2:11, 2012, https://vimeo.com/44733790.

16. Notes from telephone conversation between Norm White and Garnet Hertz, October 9th, 2014.

17. For more information on the Strategic Arts Initiative, see Graham Thomas Smith, "Strategic Arts Initiative (Program Notes)," *Graham Thomas Smith: Telepresence Art* (blog), May 28, 1986, https://designartscience.org/strategic-arts-initiative/.

18. Notes from telephone conversation between Norm White and Garnet Hertz, October 9th, 2014.

19. One notable and similar project that predates *Telephonic Arm-Wrestling* is the *Telephysical Phone System* by artist Tom Shannon. In 1969, Shannon conceived of the project that consisted of a fluid-filled sleeve around the handle of a telephone that users could squeeze and have their motion transmitted to the other telephone. For a brief description of the project, see Eduardo Kac, "Robotic Art Chronology," *Convergence* 7, no. 1 (Spring 2001): 87–111, https://www.ekac.org/robotichronology.html. For the patent, see Thomas D. Shannon, Tactile Communication Attachment, US Patent 3780225, filed January 3, 1972, and issued December 18, 1973, https://patft.uspto.gov/netacgi/nph-Parser?Sect2=PTO1&Sect2=HITOFF&p=1&u=/netahtml/PTO/search-bool.html&r=1&f=G&l=50&d=PALL&RefSrch=yes&Query=PN/3780225.

20. Norman T. White, "Norm the Artist, Oughtist, and Ne'er-Do-Well," accessed August 19, 2021, http://www.normill.ca/artpage.html.

21. Jack Burnham, "Systems Esthetics," *Artforum* 7, no. 1 (September 1968), 30–35.

22. Monoskop, "Nicolas Schöffer," accessed July 27, 2022, https://monoskop.org/Nicolas_Schöffer.

23. Marc Hassenzahl, "Experience Design: Technology for All the Right Reasons," *Synthesis Lectures on Human-Centered Informatics* 3, no. 1 (January 2010): 1–95, https://doi.org/10.2200/S00261ED1V01Y201003HCI008.

24. Thomas Birtchnell, *Indovation: Innovation and a Global Knowledge Economy in India* (Basingstoke, UK: Palgrave Macmillan, 2013).

25. Rishikesha Krishnan, *From Jugaad to Systematic Innovation: The Challenge for India*, (Utpreraka Foundation, 2010).

26. Marc Dusseiller, "Karkhana: Makers, Hackers or Slackers, " interview by Sabhyata Timsina, https://www.hackteria.org/wiki/images/e/ef/Makers%2C_hackers_or_slackers-converted.pdf.

27. Anthony Dunne and Fiona Raby, *Speculative Everything: Design, Fiction, and Social Dreaming* (Cambridge, MA: MIT Press, 2013); Lev Manovich, *The Language of New Media* (Cambridge, MA: MIT Press, 2002).

28. Butoliya, "Critical Jugaad," 79.

29. Butoliya, 2.

1.3 Frugality and the *Toaster Project*

1. Thomas Thwaites, *The Toaster Project: Or A Heroic Attempt to Build a Simple Electric Appliance from Scratch* (New York: Princeton Architectural Press, 2011), 169–170.

2. Thwaites, 19.

3. Thwaites, 38–39.

4. Anthony Dunne and Fiona Raby, *Speculative Everything: Design, Fiction, and Social Dreaming* (Cambridge, MA: MIT Press, 2013), 80.

5. Sandra Alfoldy, "Craftwashing," PechaKucha video presentation, 6:50, September 26, 2016, https://www.pechakucha.com/presentations/craftwashing.

6. Albert Borgmann, *Holding on to Reality: The Nature of Information at the Turn of the Millennium* (Chicago, IL: University of Chicago Press, 1999), 176.

7. Borgmann, 41.

8. Borgmann, 42.

9. Borgmann, 43.

10. Andrew Feenberg, "The Concept of Function in Critical Theory of Technology," in *Philosophy of Technology after the Empirical Turn*, ed. Maarten Franssen et al., Philosophy of Engineering and Technology, vol. 23 (Cham, ZG: Springer International Publishing, 2016), 283–303, https://doi.org/10.1007/978-3-319-33717-3_16.

11. Albert Borgmann, *Technology and the Character of Contemporary Life: A Philosophical Inquiry*, (Chicago, IL: University of Chicago Press, 2006), 197.

12. Borgmann, *Technology and the Character of Contemporary Life*, 197.

13. Anthony Dunne, *Hertzian Tales: Electronic Products, Aesthetic Experience, and Critical Design*, 2005 ed. (Cambridge, MA: MIT Press, 2005), 72.

14. Thwaites, *The Toaster Project*, 13.

15. Borgmann, *Technology and the Character of Contemporary Life*, 222.

16. Borgmann, 222.

17. Jérémy Bonvoisin, Robert Mies, Jean-François Boujut, Rainer Stark, "What Is the 'Source' of Open Source Hardware?," *Journal of Open Hardware* 1(1): 1–18. DOI: https://doi.org/10.5334/joh.7

18. "Marcin Jakubowski," Founder, Open Source Ecology, accessed August 19, 2021, https://www.opensourceecology.org/marcin-jakubowski/.

19. "Marcin Jakubowski."

20. Robert King Merton, *Social Theory and Social Structure* (Glencoe, IL: Free Press, 1957), 198.

21. "Marcin Jakubowski," Open Source Ecology.

Theme 2

1. Srinivas Nidhra, "Black Box and White Box Testing Techniques—A Literature Review," *International Journal of Embedded Systems and Applications* 2, no. 2 (June 30, 2012): 29–50, https://doi.org/10.5121/ijesa.2012.2204.

2. Bruno Latour, *Pandora's Hope: Essays on the Reality of Science Studies* (Cambridge, MA: Harvard University Press, 1999), 304.

3. "ESP32," Overview, Products, Espressif Systems, 2020, https://www.espressif.com/en/products/socs/esp32.

4. Michel Callon, "Techno-Economic Networks and Irreversibility," *The Sociological Review* 38, no. 1_suppl (May 1990): 132–61, https://doi.org/10.1111/j.1467-954X.1990.tb03351.x.

5. John Law, "Notes on the Theory of the Actor-Network: Ordering, Strategy, and Heterogeneity," *Systems Practice* 5, no. 4 (August 1992): 379–93, https://doi.org/10.1007/BF01059830.

6. "Collection Online: Nam June Paik," The Guggenheim Museums and Foundation, accessed August 27, 2021, https://www.guggenheim.org/artwork/artist/nam-june-paik.

7. Edith Decker-Phillips, *Paik Video* (Barrytown, NY: Barrytown, Ltd, 1998).

2.1 Exploration and the *Incantor*

1. Punctualization refers to a concept in Actor-network theory to describe when components are brought together into a single complex system that can be used as a single object. For more information, see Bruno Latour, *Pandora's Hope: Essays on the Reality of Science Studies* (Cambridge, MA: Harvard University Press, 1999).

2. For a relevant discussion of infrastructure, see Susan Leigh Star and Karen Ruhleder, "Steps Toward an Ecology of Infrastructure: Design and Access for Large Information Spaces," *Information Systems Research* 7, no. 1 (1996): 111–34; and Bernard London, *Ending the Depression through Planned Obsolescence* (New York, 1932), https://catalog.hathitrust.org/Record/006829435.

3. Qubais Reed Ghazala, "The Folk Music of Chance Electronics: Circuit-Bending the Modern Coconut," *Leonardo Music Journal* 14 (December 1, 2004): 97–104, https://doi.org/10.1162/0961121043067271.

4. James Pritchett, *The Music of John Cage* (Cambridge, UK: Cambridge University Press, 1993), 94.

5. Edith Decker-Phillips, *Paik Video* (Barrytown, NY: Barrytown, Ltd, 1998).

6. Ghazala, "The Folk Music of Chance Electronics," 97–104.

7. Texas Instruments, "TI Talking Learning Aid Sets Pace for Innovative CES Introductions," press release, June 11, 1978, http://www.datamath.org/Story/TI_PRESS.htm.

8. Reed Ghazala, "Incantors," Anti-Theory, accessed August 19, 2021, http://www.anti-theory.com/sales/sales_gallery/c/main.html.

9. Ghazala, "The Folk Music of Chance Electronics," 97–104.

10. Walter Benjamin, *The Work of Art in the Age of Mechanical Reproduction* (London: Penguin Books, 2008), 220.

11. Benjamin, *The Work of Art*, 220.

12. "Bent Festival 2009," archived April 20, 2009, https://web.archive.org/web/20090420052844/http:/www.bentfestival.org/.

13. Jeff Morton, "A Critical Perspective on Circuit-Bending," *BlackFlash*, May 16, 2013, https://blackflash.ca/2013/05/16/circuitbendingjeffmorton/.

14. Anne Galloway, Jonah Brucker-Cohen, Lalya Gaye, and Elizabeth Goodman, "Design for Hackability," in *DIS '04:Proceedings of the 5th Conference on Designing Interactive Systems: Processes, Practices, Methods, and Techniques* (New York: ACM Press, 2004), 364, https://doi.org/10.1145/1013115.1013181.

15. Amy Spencer, *DIY: The Rise of Lo-Fi Culture*, 2nd ed. (London: Marion Boyars, 2008), 11.

16. Ellen Lupton, ed., *D.I.Y.: Design It Yourself*, Design Briefs (New York: Princeton Architectural Press, 2006), 18.

17. Galloway et al., "Design for Hackability," 364.

18. Matthew Gault and Jason Koebler, "Elizabeth Warren Calls for a National Right-to-Repair Law for Tractors," *Vice*, March 27, 2019, accessed August 20, 2021, https://www.vice.com/en/article/d3mb5k/elizabeth-warren-calls-for-a-national-right-to-repair-law.

19. Latour, *Pandora's Hope*, 183.

20. For an analysis of software End User License Agreements, see Cory Doctorow, "Shrinkwrap Licenses: An Epidemic Of Lawsuits Waiting To Happen," *InformationWeek*, February 2, 2007, https://www.informationweek.com/software/shrinkwrap-licenses-an-epidemic-of-lawsuits-waiting-to-happen.

21. Ed van Hinte, et al., eds., *Young Designers Trigger Industries Trigger Young Designers* (Amsterdam: De Balie, 1996), 226–227.

22. Initially published by O'Reilly Press starting in 2005, *Make* spun off into its own corporation as Maker Media Inc. in 2013. For more information, see "Make: Community," Make: Community, accessed August 20, 2021, https://make.co.

23. In Heidegger's terminology, the ready-to-hand is a tool that is used intuitively and as an extension of oneself without being focused on as an object, while an object that is present-at-hand is objectively present and requires the user to think about it as an object. Martin Heidegger, *Being in Time*, trans. John Macquarrie and Edward Robinson (Oxford, UK: Blackwell Publishers Ltd., 1962), 99. http://pdf-objects.com/files /Heidegger-Martin-Being-and-Time-trans.-Macquarrie-Robinson-Blackwell-1962.pdf.

24. Eugene Thacker and Alexander R. Galloway, introduction to *Protocol: How Control Exists after Decentralization*, Leonardo (Cambridge, MA: MIT Press, 2004), xiii.

25. Michel de Certeau, *The Practice of Everyday Life. 1*, trans. Steven Rendall, (Berkeley, CA.: Univ. of California Press, 2013), xiv.

26. John Scanlan, *On Garbage*, repr. (London: Reaktion Books, 2005), 14–16; and Greg Siegel, *Forensic Media: Reconstructing Accidents in Accelerated Modernity* (Durham, NC: Duke University Press, 2014).

27. Emit Snake-Beings, "The DiY ['Do It Yourself'] Ethos: A Participatory Culture of Material Engagement" (PhD diss., University of Waikato, 2016), 12, https://hdl .handle.net/10289/9973.

28. Siegel, *Forensic Media*.

2.2 Exploration and *Wire Figures*

1. Norbert Wiener, *Cybernetics: Or, Control and Communication in the Animal and the Machine*, 2nd edition, 2019 reissue (Cambridge, MA: MIT Press, 2019), xi.

2. Wiener proposes that "the many automata of the present age are coupled to the outside world both for the reception of impressions and for the performance of actions. They contain sense-organs, effectors, and the equivalent of a nervous system to integrate the transfer of information from the one to the other. They lend themselves very well to description in physiological terms. It is scarcely a miracle that they can be subsumed under one theory with the mechanisms of physiology." Wiener, *Cybernetics*, 43.

3. Edward A. Shanken and Ian Farr, eds., *Systems*, Whitechapel: Documents of Contemporary Art (Cambridge, MA: MIT Press, 2015).

4. Phivos-Angelos Kollias, "Iannis Xenakis and Systems Thinking" (Paper, Xenakis International Symposium, Southbank Centre, London, April 1–3, 2011).

5. The field of artificial intelligence originated at a workshop at Dartmouth College in 1956 with Allen Newell, Herbert Simon, John McCarthy, Marvin Minsky, and Arthur Samuel. McCarthy coined the term "artificial intelligence" to splinter it away from the field of cybernetics. John McCarthy, "Review of the Question of Artificial Intelligence," *Annals of the History of Computing* 10, no. 3 (1988): 224–229.

6. Jack Burnham, "System Esthetics," *Artforum* 7, no. 1 (September 1968).

7. The first of Mims's books published by Radio Shack was *Introduction to Electronics* (1972) and the last was *Sun and Sky Monitoring Station* (2003).

8. Diana Burgoyne, "An Interview with Diana Burgoyne about Digital Bodies and Cybernetic Spaces—conducted by Jeremy Turner via email in December," 2003, https://web.archive.org/web/20111126210738/http:/openspace.ca/files/Diana%20Burgoyne%20Interview%20by%20Jeremy%20Turner,%202006(1).pdf.

9. See Paik's *K-456* in chapter 4.1 of this book.

10. Forrest M. Mims, *Introduction to Electronics*, (Radio Shack, 1972).

11. Daniel Jolliffe, *Diana Burgoyne: Gesture and the Handmade*, 2004, mixed media sculpture, 77.2 x 101.3 x 14.5 cm, Surrey Art Gallery, BC.

12. Ken Gregory, "Electronic Mail Cards: Sound Interventions Through the Postal System," Tactical, Works, Accessed November 5, 2021, http://www.cheapmeat.net/EMC.html.

13. Sarah Yates, *Beyond Electronics: A Series of Three Exhibitions Involving 11 Artists* (Preliminary Exhibition Catalog, University of Manitoba, 1989), 12.

14. Shanken and Farr, *Systems*.

15. Frank Popper, *Origins and Development of Kinetic Art*, trans. Stephen Bann, illustrated ed. (London: Studio Vista, 1968).

16. Frank Popper, *Art: Action and Participation*, trans. Stephen Bann (New York: New York University Press, 1975), 11.

17. Varvara Guljajeva, "From Interaction to Post-participation: the Disappearing Role of the Active Participant," *25th International Symposium on Electronic Art* (Gwangju, South Korea: ISEA, 2018), 92–98.

18. Shoshana Zuboff, *The Age of Surveillance Capitalism: The Fight for a Human Future at the New Frontier of Power* (New York: PublicAffairs, 2019).

2.3 Exploration and *20 Oscillators in 20 Minutes*

1. Elliot Williams, "32C3: *20 Oscillators in 20 Minutes*," *Hackaday* (blog), January 3, 2016, https://hackaday.com/2016/01/03/32c3-20-oscillators-in-20-minutes/.

2. Paper Tiger Television, "Tetsuo Kogawa Cooks Up an FM Transmitter," Art, Video Archives, 1992, https://papertiger.org/tetsuo-kogawa-cooks-up-an-fm-transmitter/.

3. John Bowers and Vanessa Yaremchuk, "The Priority of the Component, or in Praise of Capricious Circuitry," *Leonardo Music Journal* 17 (December 2007): 39–40, https://doi.org/10.1162/lmj.2007.17.39.

4. Derek Holzer, "Schematic as Score: Uses and Abuses of the (In)Deterministic Possibilities of Sound Technology," *Vague Terrain 19*, May 2, 2011, archived January 12, 2013, https://web.archive.org/web/20130112220646/http:/vagueterrain.net/journal19.

5. Katherine Moriwaki, Jonah Brucker-Cohen, Joe Saavedra, Liz Taylor, Louisa Campbell, and Liza Stark, "Scrapyard Challenge Jr., Adapting an Art and Design Workshop to Support STEM to STEAM Learning Experiences," in *Proceedings of IEEE Integrated STEM Education Conference (ISEC 2012)* (Ewing, NJ, IEEE: March 9, 2012), https:// DOI:10.1109/ISECon.2012.6204175

6. Alex G.R. McLean, Dave Griffiths, Nick Collins, and Geraint Wiggins, "Visualisation of Live Code," in *Proceedings of Electronic Visualization and the Arts Conference (EVA 2010)* (London, UK: BCS Learning and Development Ltd., 2010), https://doi.org/10.14236/ewic/EVA2010.6.

7. See https://www.jackstraw.org/youth_project/joel-ong-observe-wind/.

8. Garnet Hertz, "I'm interested in names of artists or hackers that perform live hardware development," Facebook, June 21, 2019, https://m.facebook.com/story.php?story_fbid=10106911554970291&id=6003191.

9. Afrorack (Bamanya Brian), "Africa's First Home Made Modular Synth," YouTube video, 9:48, April 2, 2018, https://www.youtube.com/watch?v=ov2sEfUv1XY.

10. "Locksport," LockWiki, last modified July 9, 2021, http://lockwiki.com/index.php/Locksport.

11. Rachel Maines, *Hedonizing Technologies: Paths to Pleasure in Hobbies and Leisure* (Baltimore, MD: Johns Hopkins University Press, 2009).

12. Garnet Hertz and Jussi Parikka, "Zombie Media: Circuit Bending Media Archaeology into an Art Method," *Leonardo* 45, no. 5 (October 2012): 424–430, https://doi.org/10.1162/LEON_a_00438.

13. Nicolas Collins, *Handmade Electronic Music: The Art of Hardware Hacking*, 2nd ed. (New York: Routledge, 2009).

14. Derek Holzer, "Sound: Schematic as Score (Uses of Failure)," *Word Object* (blog), May 8, 2011, https://wordobject.wordpress.com/2011/05/08/sound-schematic-as-score/.

15. Ableton AG "Live: What is Live," accessed July 27, 2022, https://www.ableton.com/en/live/what-is-live/.

16. Holzer, "Sound."

17. Holzer.

18. Darsha Hewitt, *"20 Oscillators in 20 Minutes,"* Works, accessed August 27, 2021, https://darsha.org/artwork/20-oscillators-in-20-minutes/.

Theme 3

1. Amy Spencer, *DIY: The Rise of Lo-Fi Culture*, rev. ed. (London: Marion Boyars Publishers Ltd., 2008), 11.

2. Elizabeth A. Grosz, *Volatile Bodies: Toward a Corporeal Feminism*, Theories of Representation and Difference (Bloomington: Indiana University Press, 1994), 14.

3. Linda Martín Alcoff, *Visible Identities* (Oxford, UK: Oxford University Press, 2006), https://doi.org/10.1093/0195137345.001.0001.

3.1 Identity and *Taratter MI-03*

1. 明和電機—Maywa Denki, "社史 Company History," accessed August 22, 2021, https://www.maywadenki.com/history/.

2. "Otamatone," 21st Century Digital Art, May 22, http://www.digiart21.org/art/otamatone.

3. 明和電機—Maywa Denki, タラッター / *Taratter* (1993) custom electric-powered musical instrument. https://www.maywadenki.com/products/tsukuba/taratter/.

4. Steven M. Johnson, *Patent Depending: A Collection*, ill. ed. (Steven M. Johnson, 2017).

5. Kenji Kawakami and Dan Papia, *99 More Unuseless Japanese Inventions: The Art of Chindogu*. American ed. (New York: W. W. Norton & Co, 1998).

6. Compare this to Krzysztof Wodiczko's concept of Interrogative Design, which is also an intentionally problematic type of industrial design.

7. The product was a grand failure for many reasons, including apparently that some people don't want to look like Steve Mann.

8. Jōichi Itō, ed., *Resisting Reduction: Designing Our Complex Future with Machines* (Cambridge, MA: MIT Press, 2019); and Machiko Kusahara, "Device Art: A New Form of Media Art from a Japanese Perspective," *Intelligent Agent* 6, no. 2 (2006): 4.

9. Kusahara, "Device Art," 4.

10. Kusahara, "Device Art."

11. Machiko Kusahara, "Bridging Art, Technology, and Pop Culture: Some Aspects of Japanese New Media Art Today," in *The Routledge Handbook of New Media in Asia*, ed. Larissa Hjorth and Olivia Khoo (London: Routledge, Taylor & Francis Group, 2016), 67.

12. James Auger, "Alternative Presents and Speculative Futures: Designing Fictions through the Extrapolation and Evasion of Product Lineages," *Negotiating Futures—Design Fiction* 6 (2010): 42–57.

13. Kaiwen Yang, "Counterfactual Histories and Speculative Immersion: Exploring the Capabilities of Unveiling Historical Wrongs and Institutional Discrimination" (master's thesis, Emily Carr University of Art and Design, 2021), https://ecuad.arcabc.ca/islandora/object/ecuad%3A16883?solr_nav%5Bid%5D=6c09a66b9a6934b9102a&solr_nav%5Bpage%5D=0&solr_nav%5Boffset%5D=0.

14. Amber Jamieson, "Meet the Queen of Useless Robots: 'The Internet Is a Weird Place,'" *The Guardian*, March 17, 2016, http://www.theguardian.com/technology/2016/mar/17/youtube-queen-robots-simone-giertz-inventor.

15. Simone Giertz, "I Made a Lipstick Robot," YouTube video, 0:06, February 5, 2016, https://www.youtube.com/watch?v=WcW70-6eQcY.

16. Maywa Denki and Minoru Hatanaka, *Maywa Denki: The Nonsense Machines* (Tokyo: NTT Publishing, 2004).

3.2 Identity and the *Barbie Liberation Organization*

1. Andrew McClary, *Good Toys, Bad Toys: How Safety, Society, Politics, and Fashion Have Reshaped Children's Playthings* (Jefferson, NC: McFarland & Co, 2004), 34–35.

2. Associated Press, "Company News: Mattel Says It Erred; Teen Talk Barbie Turns Silent on Math," *New York Times*, October 21, 1992, Business, https://www.nytimes.com/1992/10/21/business/company-news-mattel-says-it-erred-teen-talk-barbie-turns-silent-on-math.html.

3. The Teen Talk Barbie voice circuit included a random assortment of four phrases out of 270 possibilities when it was released in July 1992. After the "Math class is hard" phrase was deleted in October 1992, the toy included four random phrases out of 269 possibilities. According to product packaging, the "Electronic Battle Command Duke" G. I. Joe could say thousands of battle commands and sounds.

4. Vincent Bonin, "Igor Vamos," Biography, Daniel Langlois Foundation, 2002, https://www.fondation-langlois.org/html/e/page.php?NumPage=37.

5. The Yes Men included Igor Vamos, Jacques Servin, and Patrick Lichty.

6. Mark Dery, *Culture Jamming: Hacking, Slashing and Sniping in the Empire of Signs*, Open Magazine Pamphlet Series (Westfield, NJ: Open Magazine, 1993), https://www.markdery.com/books/culture-jamming/.

7. Kalle Lasn, "Culture Jams and Meme Warfare: Kalle Lasn, Adbusters, and Media Activism," interview by Wendi Pickerel, Helena Jorgensen, and Lance Bennett, 2002, 18. Accessed July 27, 2022. https://depts.washington.edu/gcp/pdf/culturejamsandmemewarfare.pdf.

8. Billboard Liberation Front & Friends, *The Art & Science of Billboard Improvement* (San Francisco, CA: Los Cabrones Press, 1990), https://toleratedindividuality.files.wordpress.com/2015/02/art-and-science-of-billboard-improvement-2nd-edition.pdf.

9. Billboard Liberation Front, *Art & Science of Billboard Improvement.*

10. Billboard Liberation Front, "McDonald's: I'm Sick of It," accessed August 24, 2021, http://www.billboardliberation.com/2010/06/06/mcdonalds-im-sick-of-it/.

11. Judy Wajcman, *Feminism Confronts Technology* (Cambridge, UK: Polity Press, 1991), 49, https://www.politybooks.com/bookdetail?book_slug=feminism-confronts-technology--9780745607788.

12. Anne Marie Balsamo, *Designing Culture: The Technological Imagination at Work* (Durham, NC: Duke University Press, 2011).

13. Balsamo, *Designing Culture*, 3.

14. Scott Eberle, "April Fools! Remember the Barbie Liberation Organization!," (blog), The Strong, April 1, 2013, https://www.museumofplay.org/blog/play-stuff/2013/04/april-fools-remember-the-barbie-liberation-organization.

15. Guy Debord and Gil Wolman, "A User's Guide to Détournement," Bureau of Public Secrets, 2007, http://www.bopsecrets.org/SI/detourn/htm.

16. Igor Vamos, "Barbie Liberation Organization," in *Beautiful Trouble: A Toolbox for Revolution*, ed. Andrew Boyd and Dave Oswald Mitchell (New York: OR Books, 2012), 282–285.

17. Vamos, "Barbie Liberation Organization."

18. Featured on News 8 San Diego, Hannah's statement aired as a local news story about the Barbie Liberation Organization project. It is unclear as to whether Henzey was recruited by the BLO before the interview.

19. David Firestone, "While Barbie Talks Tough, G. I. Joe Goes Shopping," *New York Times*, December 31, 1993, US, https://www.nytimes.com/1993/12/31/us/while-barbie-talks-tough-g-i-joe-goes-shopping.html.

20. Alan Gilbert, "Anne Waldman Changing the Frequency," *Jacket 27*, April 2005, http://jacketmagazine.com/27/w-gilb.html.

21. Suzann Kingston summarizes a critique of Barbie as follows: "Today our obsession with the Barbie look just makes me mad and sad. Why? Because it's just one version of what beauty can be—not the only one. . . . Despite the fact that the real world offers up a glorious smorgasbord of beauty for our consideration, many of us continue to buy into the 'one size fits all' version that deliberately ignores age, ethnicity, body type, genetics, etc. Strong is the woman who can confidently embrace how she looks in this society if she bears no resemblance to Barbie because the media keeps ramming that look down our collective throats." Suzann Kingston, *Step Aside, Barbie! You're Blocking My View—of Me! Women and Body Image in North America* (Vancouver, BC: Shasuze Books, 2006), 6.

22. Vamos, "Barbie Liberation Organization," 282–285.

3.3 Identity and the *Stock Market Skirt*

1. Nancy Paterson, interview by Ann Dean and Luis Jacob, February 16, 2014, raw transcript, http://www.vacuumwoman.com/About/Main/1-Transcription-final-Feb16-14.doc.

2. Nancy Paterson, "Cyberfeminism," in *Mediaworks by Nancy Paterson*, The Art Gallery of Peterborough, exhibition catalog (Peterborough, ON: The Gallery, 1995), 4. For a thorough and interesting history of the concept of cyberfeminism, see Susanna Paasonen, "Revisiting Cyberfeminism," *Communications* 36(3): 335–352.

3. Paterson, "Cyberfeminism," 48–55.

4. Jennifer S. Light, "When Computers Were Women," *Technology and Culture* 40, no. 3 (1999): 455–83.

5. Paterson, "Cyberfeminism," 48.

6. Randy Lee Cutler, "Remapping the Terrain," in *Mediaworks: Nancy Paterson*, Surrey Art Gallery, exhibition catalog (Surrey, BC: Surrey Art Gallery, 2001), 23.

7. Desmond Morris, *Manwatching: A Field Guide to Human Behaviour* (London: Elsevier, 1977), 329–333.

8. Nancy Paterson, Liane Davidson, Randy Lee Cutler, and Carol Gigliotti, *Media Works: Nancy Paterson*, Surrey Art Gallery, exhibition catalog (Surrey, BC: Surrey Art Gallery, 2001).

9. Georg Simmel, "Fashion," *American Journal of Sociology* 62, no. 6 (1957): 541–558.

10. Other early Internet-controlled electronic pieces include *CyberSM III* by Kirk Wolfond and Stahl Stenslie, *Where I Can See My House from Here So We Are* by Ken Feingold, Stelarc's *Fractal Flesh: Internet Body*, Eduardo Kac's *Rara Avis* from 1996, Stelarc's *Ping Body*, *Telegarden* by Ken Goldberg, Joseph Santarromana, and other

collaborators (June 1995), Garnet Hertz's *Interface* (April 1996), *Legal Tender* by Ken Goldberg, Eric Paulos, John Canny, Judith Donath, and Mark Pauline (1996), and Masaki Fujihata's *Light on the Net* (1996).

11. Ken Goldberg, Steve Gentner, Carl Sutter, and Jeff Wiegley, "The Mercury Project: A Feasibility Study for Internet Robots," *IEEE Robotics & Automation Magazine* 7, no. 1 (1999): 35–40.

12. From a technical perspective, the system runs a Perl script on an open-source Linux PC, which continuously monitors stock prices or composite indexes and sends control data to a stepper motor each second. When financial markets are closed, the system retrieves data from open markets in other time zones or replays historical data.

13. Cutler, "Remapping the Terrain," 23.

14. Ken Goldberg, ed., *The Robot in the Garden: Telerobotics and Telepistemology in the Age of the Internet*, Leonardo (Cambridge, MA: MIT Press, 2000).

15. Eduardo Kac, "Telepresence Art," in *Teleskulptur*, ed. Richard Kriesche (Graz, AT: Kulturdata, 1993), 48–72, http://www.ekac.org/telepresence.art._94.html.

16. danah boyd and Kate Crawford, "Critical Questions for Big Data: Provocations for a Cultural, Technological, and Scholarly Phenomenon," *Information, Communication & Society* 15, no. 5 (June 2012): 662–79, https://doi.org/10.1080/1369118X.2012.678878.

17. Marshall McLuhan, *Understanding Media: The Extensions of Man* (Canada: McGraw-Hill, 1964).

18. Marshall McLuhan, "Part III, 13," in *Report on Project in Understanding New Media* (National Association of Educational Broadcasters, 1960).

19. Marshall McLuhan and Quentin Fiore, *The Medium Is the Massage* (New York: Random House, 1967), 40.

20. Sadie Plant, *Zeros + Ones: Digital Women + the New Technoculture* (London: Doubleday, 1997), 37.

21. Diana Crane, *Fashion and Its Social Agendas: Class, Gender, and Identity in Clothing* (Chicago, IL: University of Chicago Press, 2000), 1.

22. Anne Balsamo, "Forms of Technological Embodiment: Reading the Body in Contemporary Culture," *Body & Society* 1, no. 3–4 (November 1995): 215, https://doi.org/10.1177/1357034X95001003013.

23. Balsamo, "Forms of Technological Embodiment," 217.

24. Howard Rheingold, *Virtual Reality* (New York: Summit Books, 1991).

25. Sherry Turkle, *Life on the Screen: Identity in the Age of the Internet* (New York: Simon and Schuster, 1995), 184.

26. Michael Cavna, "Nobody Knows You're a Dog: As Iconic Internet Cartoon Turns 20, Creator Peter Steiner Knows the Joke Rings as Relevant as Ever," *The Washington Post*, July 31, 2013. https://www.washingtonpost.com/blogs/comic-riffs/post/nobody-knows-youre-a-dog-as-iconic-internet-cartoon-turns-20-creator-peter-steiner-knows-the-joke-rings-as-relevant-as-ever/2013/07/31/73372600-f98d-11e2-8e84-c56731a202fb_blog.html

27. Balsamo, "Forms of Technological Embodiment," 233.

28. Balsamo, 215.

29. Donna Haraway, "A Cyborg Manifesto: Science, Technology, and Socialist-Feminism in the Late Twentieth Century," chap. 8 in *Simians, Cyborgs, and Women: The Reinvention of Nature* (New York: Routledge, 2013).

30. Donna Jeanne Haraway, *The Companion Species Manifesto: Dogs, People, and Significant Otherness*, Paradigm 8 (Chicago, IL: Prickly Paradigm Press, 2003), 4.

31. Balsamo, "Forms of Technological Embodiment," 215.

32. Hélène Day Fraser and Keith Doyle, "Clothing as Conversation: Conversations," accessed August 12, 2022, http://clothingasconversation.com/conversations/.

33. See Hélène Day Fraser, Keith Doyle, Nicolene McKenzie, Mia Daniels, Natalie Tillen, "Yes, And (+), Maybe: cloTHING(s) as Conversation," *Cumulus 2016 Conference Proceedings*, April 27–May 1, 2016, 36–42. Bill. Gaver, Tony Dunne, and Elena Pacenti, "Design: Cultural Probes," *Interactions* 6, no. 1 (1999), 21–29. Elizabeth B. N. Sanders, and Pieter Jan Stappers, "Co-creation and the New Landscapes of Design," *Co-design* 4, no. 1 (2008), 5–18. Anthony Dunne and Fiona Raby, *Speculative Everything: Design, Fiction, and Social Dreaming* (Cambridge, MA: MIT Press, 2013). Dave Hiles, "Heuristic Inquiry and Transpersonal Research," *Annual Meeting of the Centre for Counselling and Psychotherapy Education*, no. 25 (2007); Valerie Malhotra Bentz and J. Jeremy Shapiro, *Mindful Inquiry in Social Research* (Thousand Oaks, CA: Sage Publications, 1998).

34. Kat Jungnickel, *Bikes and Bloomers: Victorian Women Inventors and Their Extraordinary Cycle Wear* (Cambridge, MA: Goldsmiths Press, 2018), 121–154.

35. Nancy Paterson, "IXmaps," accessed August 22, 2021, http://www.vacuumwoman.com/About/Main/ixmaps.html.

36. Mark Jones, "Cyberfeminist at Large: Nancy Paterson Raises Her Skirt," *CyberStage* 3, no. 1 (1998).

37. Anne Balsamo, *Designing Culture: The Technological Imagination at Work* (Durham, NC: Duke University Press, 2011), 32, https://doi.org/10.1215/9780822392149.

Theme 4

1. Catherine Flood and Gavin Grindon, *Disobedient Objects* (London: Victoria & Albert Museum, 2014).

2. An archive of the downloadable instructions from the V&A exhibition, which ran from July 26, 2014 to February 1, 2015, can be seen at: "Disobedient Objects: How-To Guides," Victoria and Albert Museum, http://www.vam.ac.uk/content/exhibitions/disobedient-objects/how-to-guides/.

3. "Disobedient Objects: How-To Guides."

4. Catherine Flood and Gavin Grindon, *Disobedient Objects* (London: V & A Publishing, 2014), 9.

5. Julian Oliver, Gordon Savičić, and Danja Vasiliev, "The Critical Engineering Manifesto," *The Critical Engineering Working Group* 2016 (2011), http://criticalengineering.org/ (retrieved September 2, 2022); Matt Ratto, "Critical Making: Conceptual and Material Studies in Technology and Social Life," *The Information Society* 27, no. 4 (2011): 252–260; Garnet Hertz, ed., *Critical Making*, Telharmonium Press, 2012; Phil Agre, "Computing as a Social Practice," in *Reinventing Technology, Rediscovering Community*, ed. P. Agre and D. Schuler (Santa Barbara:, CA: Praeger, 1997); Phoebe Sengers, Kirsten Boehner, Shay David, and Joseph "Jofish" Kaye, "Reflective Design," *Proceedings of the 4th Decennial Conference on Critical Computing: Between Sense and Sensibility* (Aarhus, Denmark: 2005), 49–58; Anthony Dunne, *Hertzian tales: Electronic Products, Aesthetic Experience, and Critical Design* (Cambridge, MA: MIT Press, 2008); Carl DiSalvo, *Adversarial Design* (Cambridge, MA: MIT Press, 2015); Krsysztof Wodiczko, *Critical Vehicles: Writings, Projects, Interviews* (Cambridge, MA: MIT Press, 1999).

6. Wodiczko, *Critical Vehicles*.

7. David Garcia and Geert Lovink, "The ABC of Tactical Media," May 16, 1997, https://www.nettime.org/Lists-Archives/nettime-l-9705/msg00096.html.

8. Anthony Dunne and Fiona Raby, *Speculative Everything: Design, Fiction, and Social Dreaming* (Cambridge, MA: MIT Press, 2013).

9. "Interrogative," Interrogative Design Group, MIT, archived June 22, 2010, https://web.archive.org/web/20100622044914/http:/interrogative.mit.edu/about/.

10. "Krzysztof Wodiczko," About, Krzysztof Wodiczko, accessed August 22, 2021, https://www.krzysztofwodiczko.com/about.

11. "Statement of Purpose," Interrogative Design Group, MIT, archived April 22, 2001, http://web.archive.org/web/20010422045615fw_/http://web.mit.edu/idg/purpose.html.

12. Stephen Holden, "All Suited Up for Mischief, to Rumple Stuffed Shirts," *New York Times*, October 6, 2009, Movies, https://www.nytimes.com/2009/10/07/movies/07yes.html.

13. According to Critical Art Ensemble's *Digital Resistance: Explorations in Tactical Media* (New York: Autonomedia, 2001), 4:

> The process began in 1993 when a coalition of Dutch cultural groups produced an event/scene in Amsterdam called the Next 5 Minutes (N5M). The topic of the event was 'Tactical Television' (so named by Dutch cultural theorists involved in the production who were inspired by de Certeau's work, The Practice of Everyday Life). The event drew all kinds of people from Europe and North America who were concerned with issues of intervening in television, theorizing the structure and dynamics of video culture, modelling representations of political causes that further social justice, creating alternative models of distribution, and so on. The event was small (around three hundred people), but it indicated that a new kind of coalition was beginning to form. Event organizers quickly realized that tactical television was too limited in its scope, because there were people with a similar sensibility who were doing tactical work in all types of media, and that they should all come together. The event's next manifestation in 1996 addressed the topic 'Tactical Media' (as it did again in 1999).

14. John G. Hanhardt, Nam June Paik, and Solomon R. Guggenheim Museum, *The Worlds of Nam June Paik* (New York: Guggenheim Museum Publications, 2000), 100.

15. Michael Shamberg, *Guerrilla Television* (New York: Holt, Rinehart, Winston, 1973).

16. "Tactical Media Crew," English Language Home page," accessed August 22, 2021, https://www.tmcrew.org/enghome.htm.

17. Critical Art Ensemble, *Digital Resistance*, 5.

18. Critical Art Ensemble, 5.

19. Michel de Certeau, *The Practice of Everyday Life*, trans. Steven Rendall (Berkeley, CA: University of California Press, 2013).

20. David Garcia and Geert Lovink, "The ABC of Tactical Media," May 16, 1997, https://www.nettime.org/Lists-Archives/nettime-l-9705/msg00096.html.

21. Rita Raley, *Tactical Media*, NED-New edition, vol. 28 (Minneapolis: University of Minnesota Press, 2009), 12, https://www.jstor.org/stable/10.5749/j.ctttsbc0.

22. Rita Raley, *Tactical Media*, 11.

23. Rita Raley, *Tactical Media*.

24. Rita Raley, 4.

25. Alexander R. Galloway & Eugene Thacker, *The Exploit, A Theory of Networks, Electronic Mediations*, vol. 21 (Minneapolis: University of Minnesota Press, 2007), 81.

26. Barbara Abrash & Faye Ginsburg (eds), "Defining Tactical Media," *The Virtual Casebook Project*, Archived March 25, 2016, https://web.archive.org/web/20160325023532/http://www.nyu.edu/fas/projects/vcb/definingTM.html.

27. Thomas Snow, "Review of 'Disobedient Objects,'" *Object* 17, no. 1 (December 3, 2015): 128-130, https://student-journals.ucl.ac.uk/obj/article/id/338/.

4.1 Disobedience and *Robot K-456*

1. Dick Hebdige, *Subculture: The Meaning of Style* (London: Routledge, 1979).

2. It should be clearly noted that the punk movement did not start and end with the Sex Pistols. See John Lydon, "A Beautiful Ugliness Inside," in *Punk: Chaos to Couture*, Andrew Bolton, Richard Hell, John Savage, and John Lydon (New York: Metropolitan Museum of Art, 2013), 21.

3. John G. Hanhardt, Nam June Paik, and Solomon R. Guggenheim Museum, *The Worlds of Nam June Paik* (New York: Guggenheim Museum Pubns, 2000).

4. For a detailed bibliography on Paik, see Nam June Paik, Wulf Herzogenrath, and Sabine Maria Schmidt, *Nam June Paik: Fluxus, Video* (Bremen, DE: Kunsthalle Bremen, 1999). For postmillennial sources, see Nam June Paik, Sook-Kyung Lee, and Susanne Rennert, *Nam June Paik* (London: Tate, 2010), 230–31.

5. Edward Shanken, "Art in the Information Age: Cybernetics, Software, Telematics and the Conceptual Contributions of Art and Technology to Art History and Aesthetic Theory," (PhD diss., Duke University, 2001), 84, https://artexetra.files.wordpress.com/2010/04/shanken_art_info_age_diss_2001.pdf.

6. Nam June Paik, *Nam June Paik: Video Works, 1963–88: 29 September-11 December 1988, Hayward Gallery* (London: South Bank Centre, 1988), 34–35.

7. Jack Burnham, *Beyond Modern Sculpture: The Effects of Science and Technology on the Sculpture of This Century* (New York: George Braziller, 1968), 351.

8. Paik, *Nam June Paik: Video Works*, 22.

9. Paik, 5.

10. "Nam June Paik," Collection Online, The Guggenheim Museums and Foundation, accessed August 23, 2021, https://www.guggenheim.org/artwork/artist/nam-june-paik.

11. Paik, *Nam June Paik: Video Works*, 20.

12. Hanhardt, Paik, and Guggenheim Museum, *The Worlds of Nam June Paik*, 57.

13. Paik, Lee, and Rennert, *Nam June Paik*, 162.

14. Mozart's *K-456* is commonly referred to as "Piano Concerto No. 18 in B-flat major" and sometimes called the "Paradis." See Hermann Ullrich, "Maria Theresia Paradis and Mozart," *Music and Letters* 27, no. 4 (October 1, 1946): 224–33, https://doi.org/10.1093/ml/XXVII.4.224.

15. Joan Rothfuss and Yōko Ono, *Topless Cellist: The Improbable Life of Charlotte Moorman* (Cambridge, MA: MIT Press, 2014), 86.

16. Paik, *Nam June Paik*, 31.

17. Paik, *Nam June Paik: Video Works*, 21.

18. Paik, *Nam June Paik*, 12.

19. John G. Hanhardt, "Passages: Nam June Paik," *Artforum*, April 2006, https://www.artforum.com/print/200604/nam-june-paik-10623.

20. "Nam June Paik Exhibition, Extraordinary Phenomenon," Exhibition, Nam June Paik Art Center, 2017, archived January 4, 2019, https://web.archive.org/web/20190104045017/https://njpac-en.ggcf.kr/archives/exhibit/extraordinary-extraordinary-phenomenon-nam-june-paik.

21. Barbara Maria Stafford, Frances Terpak, and Isotta Poggi, *Devices of Wonder: From the World in a Box to Images on a Screen* (Los Angeles: Getty Research Institute, 2001.

22. Melissa Chiu and Michelle Yun, eds., "Ok, Let's Go to Blimpies: Talking about Nam June Paik," *Nam June Paik: Becoming Robot*, ill. ed. (New York: The Asia Society Museum, 2014), 40–41.

23. Jack Burnham, "Robot and Cyborg Art," in *Beyond Modern Sculpture* (New York: George Braziller, 1968), 351. *K-456* used parts of Paik's nephew's (Ken Paik Hakuta) toys, and cups from his sister's bra. Blair Asbury Brooks, "How Nam June Paik Used Technology to Search for a Deeper Humanity," *Artspace*, October 31, 2014, http://www.artspace.com/magazine/art_101/close-look-nam-june-paik.

24. Everyday items were in common use far before *K-456*—for example, Picasso's collage *Guitar* (1913) included a front page of a printed newspaper glued onto the surface—but specifically referring to them as "poor" was not common. WikiArt, s.v. "*The Guitar*, Pablo Picasso," last modified February 12, 2013, https://www.wikiart.org/en/pablo-picasso/the-guitar-1913.

25. Silvia Lindtner, Shaowen Bardzell, and Jeffrey Bardzell, "Reconstituting the Utopian Vision of Making: HCI After Technosolutionism," in *CHI '16: Proceedings of the 2016 CHI Conference on Human Factors in Computing Systems* (San Jose, CA: ACM, 2016), 183, https://doi.org/10.1145/2858036.2858506.

26. Evgeny Morozov, *To Save Everything, Click Here: The Folly of Technological Solutionism* (New York: PublicAffairs, 2013).

27. Paik, *Nam June Paik: Video Works*, 22.

28. Paik, *Nam June Paik: Video Works,*, 11.

29. Thomas Thwaites, *The Toaster Project: Or A Heroic Attempt to Build a Simple Electric Appliance from Scratch* (New York: Princeton Architectural Press, 2011), 169–170.

30. Leonard Koren, *Wabi-Sabi for Artists, Designers, Poets & Philosophers* (Berkeley, CA: Stone Bridge Press, 1994).

31. Taro Gold, *Living Wabi Sabi: The True Beauty of Your Life*, 2nd ed. (Kansas City, MO: Andrews McMeel Publishing, 2004), 20–21.

32. Daisetz T. Suzuki, *Zen and Japanese Culture* (Princeton, NJ: Princeton University Press, 2019), 284, https://doi.org/10.2307/j.ctvc77449.

33. Tom Jennings, "Wabi Tek Sabi," Sensitive Research, February 2020, https://www.sr-ix.com/Etc/wabiteksabi/.

34. Stephen Duncombe, *Notes from Underground: Zines and the Politics of Alternative Culture*, 2nd ed. (Bloomington, IN: Microcosm Publishing, 2008), 47.

35. Duncombe, *Notes from Underground*, 46.

36. Jean Lave and Etienne Wenger, *Situated Learning: Legitimate Peripheral Participation*, Learning in Doing (Cambridge, UK: Cambridge University Press, 1991).

37. Rafael Lozano-Hemmer, "Perverting Technological Correctness," *Leonardo* 29, no. 1 (1996): 5–15.

38. Susan Chaityn Lebovits, "Living a Wabi Sabi Life: An Interview with Robyn Griggs Lawrence," *EcoSalon*, December 10, 2008, http://ecosalon.com/living_a_wabi_sabi_life_and_tips_to_get_started/.

4.2 Disobedience and *Hairbrain 2000*

1. NHẾT TV, "Chế Tạo Lamboghini Từ Xe Đạp | Lamboghini Preparations from Bicycles | Lamborghini 100 $ Version," YouTube video, 15:08, June 23, 2019, https://www.youtube.com/watch?v=H81_W725ZS4.

2. *Brainstorm*, directed by Douglass Trumbull (Beverly Hills, CA: JF Productions, Metro-Goldwyn-Mayer, 1983), https://www.imdb.com/title/tt0085271.

3. Laua Kikauka, telephone interview by author, June 23, 2017.

4. Ivan E. Sutherland, "A Head-Mounted Three Dimensional Display," in *AFIPS '68 (Fall, part I): Proceedings of the December 9–11, 1968, Fall Joint Computer Conference* (New York: ACM Press, 1968), 757–64, https://doi.org/10.1145/1476589.1476686.

5. For example, Philco, an American electronics manufacturer based in Philadelphia, had done some work in head-mounted displays and head tracking in 1961, but this system transmitted live video from a remote location instead of showing

computer-generated images. Roy S. Kalawsky, *The Science of Virtual Reality and Virtual Environments: A Technical, Scientific and Engineering Reference on Virtual Environments* (Wokingham, UK: Addison-Wesley, 1993).

6. Andrew Sears and Julie A. Jacko, eds., *The Human-Computer Interaction Handbook: Fundamentals, Evolving Technologies and Emerging Applications,* 2nd ed. (New York: CRC Press, 2007).

7. Marcus Tullius Cicero, *On Life and Death*, ed. John Davie and Miriam T. Griffin, Oxford World's Classics (Oxford, UK: Oxford University Press, 2017).

8. Sutherland, "Three Dimensional Display," 759.

9. Sutherland, 758.

10. Andrew Leonard, "Wired: The Book," *Salon*, August 20, 1998, https://www.salon.com/1998/08/20/featureb_12/.

11. Howard Rheingold, *Virtual Reality* (New York: Summit Books, 1991).

12. Mitch Altman is best known as the inventor of the *TV-B-Gone* device, an open source hardware project for turning off public televisions. He is also cofounder of Noisebridge Hackerspace in San Francisco, one of the first North American hackerspaces. For more on Altman, see his company website at "Cornfield Electronics," Home, accessed August 22, 2021, https://cornfieldelectronics.com/cfe/cfe.main.php.

13. Myron W. Krueger, *Artificial Reality* (Reading, MA: Addison-Wesley, 1983).

14. David Jay Brown, Rebecca McClen Novick, and Jerry Garcia, eds., *Voices from the Edge: Conversations with Jerry Garcia, Ram Dass, Annie Sprinkle, Matthew Fox, Jaron Lanier, & Others* (Freedom, CA: Crossing Press, 1995).

15. G. Pascal Zachary, "Artificial Reality Computer Simulations May One Day Provide Surreal Experiences," *Wall Street Journal*, January 23, 1990, 1.

16. Genevieve Bell and Paul Dourish, "Yesterday's Tomorrows: Notes on Ubiquitous Computing's Dominant Vision," *Personal and Ubiquitous Computing* 11, no. 2 (January 24, 2007): 133–43, https://doi.org/10.1007/s00779-006-0071-x.

17. Mark Weiser, "The Computer for the 21st Century," *Scientific American* 265, no. 3 (September 1991): 94–104, https://doi.org/10.1038/scientificamerican0991-94.

18. Mary Anne Moser and Douglas MacLeod, eds., *Immersed in Technology: Art and Virtual Environments*, Leonardo (Cambridge, MA: MIT Press, 1996), xix.

19. "Meta Quest 2: Health and Safety," Meta Quest, accessed July 27, 2022, https://www.oculus.com/safety-center/quest-2/.

20. "NeoLucida," NeoLucida, accessed August 26, 2021, https://neolucida.com.

21. The Flintstones is an American cartoon from the 1960s by Hanna-Barbera that features devices like a woodpecker-powered sewing machine, a pelican-powered dishwashing machine, a brontosaurus crane, and a lizard-powered can opener.

22. Rafael Lozano-Hemmer, "Perverting Technological Correctness," *Leonardo* 29, no. 1 (February 1996): 6, https://muse.jhu.edu/article/607159.

23. Richard Barbrook and Andy Cameron, "The Californian Ideology," *Mute* 1, no. 3 (September 1995), https://www.metamute.org/editorial/articles/californian-ideology.

24. Paul Duguid, "Material Matters: The Past and the Futurology of the Book," in *The Future of the Book*, ed. G. Nunberg (Berkeley, CA: University of California Press, 1996), 65.

25. "Meta Quest 2."

4.3 Disobedience and *Feral Robotic Dogs*

1. "Feral Dogs: Site Statistics, East 173rd Street Works/Former Gas Plant," Projects, NYU, January 2003, archived November 5, 2010, https://web.archive.org/web/20101105122728/http:/www.nyu.edu/projects/xdesign/feralrobots/site_nyc_bronxriver.html.

2. "Home," Figaro USA, Inc., accessed August 24, 2021, https://www.figarosensor.com/.

3. The collaborators on the *Feral Robotic Dogs* project include Natalie Jeremijenko, Will Kavesh, Jesse Arnold, Mike Lin, and Mike Kai.

4. Karen Peart, "'Feral' Robot Dogs Trace Pedigree to Yale Engineer," *Yale Bulletin & Calendar* 31, no. 25, April 11, 2003, archived April 23, 2009, https://web.archive.org/web/20090423215758/www.yale.edu/opa/arc-ybc/v31.n25/story2.html.

5. Sanjida O'Connell, "Who Let the Robot Dogs Out?," *Independent*, December 3, 2008, https://www.independent.co.uk/life-style/gadgets-and-tech/features/who-let-the-robot-dogs-out-1048566.html.

6. Rita Raley, *Tactical Media*, NED-New edition, vol. 28 (Minneapolis: University of Minnesota Press, 2009), 9, https://www.jstor.org/stable/10.5749/j.ctttsbc0.

7. Joanne Richardson, "Language of Tactical Media," subsol, August 2002, http://subsol.c3.hu/subsol_2/contributors2/richardsontext2.html.

8. Douglas Back, "The Babbage Disease" (presentation, ISEA95: Sixth International Symposium on Electronic Art, Montreal, Canada, September 1995), https://isea-archives.siggraph.org/presentation/the-babbage-disease/.

9. David Garcia and Geert Lovink, "The ABC of Tactical Media," May 16, 1997, https://www.nettime.org/Lists-Archives/nettime-l-9705/msg00096.html.

10. Peart, "'Feral' Robot Dogs Trace Pedigree."

11. Critical Art Ensemble, *Digital Resistance: Explorations in Tactical Media* (New York: Autonomedia, 2001), 8.

12. Heterotopia is a term used by Michel Foucault to describe the opposite of utopia. Heterotopias are generally difficult, abnormal, contrasting, or transformational spaces. Heterotopias are almost a subworld within the world, echoing but also flipping the customs of standard society. Foucault gives several examples including brothels, jails, boats, graveyards, bars, and fairs. Michel Foucault, *The Order of Things: An Archaeology of the Human Sciences* (New York: Vintage Books, 1994).

13. Lisa Parks, "Falling Apart: Electronics Salvaging and the Global Media Economy," in *Residual Media,* ed. Charles R. Acland (Minneapolis: University of Minnesota Press, 2007), 32–47.

14. Benjamin Galon: http://recyclism.com/refunctmedia.html. Katherine Moriwaki and Jonah Brucker-Cohen: http://www.scrapyardchallenge.com. Ken Gregory: http://www.cheapmeat.net.

15. Bernard London, *Ending the Depression through Planned Obsolescence* (New York, 1932), https://catalog.hathitrust.org/Record/006829435.

16. Glenn Adamson and Brooks Stevens, *Industrial Strength Design: How Brooks Stevens Shaped Your World* (Cambridge, MA: MIT Press, 2003), 129.

17. Victor Lebow, "Price Competition in 1955," *Journal of Retailing* 31, no. 1 (Spring 1955): 7, https://www.academia.edu/21551517/Price_Competition_in_1955.

18. Richard Maxwell and Toby Miller, "E-Waste: Elephant in the Living Room," *Flow,* December 2, 2008, http://www.flowjournal.org/2008/12/e-waste-elephant-in-the-living-room-richard-maxwell-queens-college-cuny-toby-miller-uc-riverside/.

19. Natalie Jeremijenko, "Engineering Anti-Techno-Fetishism: Natalie Jeremijenko in Conversation with Garnet Hertz," in *Conversations in Critical Making,* ed. Garnet Hertz (Blueshift Series, 2015).

20. For example, Shih Chieh Huang's 2008 project *Twilight Zone* is a collection of jellyfish-like creations constructed primarily out of electronic parts commonly used to customize desktop gaming computers. Small microprocessor cooling fans lit with multicolored light-emitting diodes (LEDs), power bars, cables, and zip ties are stuck together in symmetrical, creature-like bodies. For more information on Shih Chieh Huang, see "Shih Chieh Huang 黃世傑," accessed August 24, 2021, http://www.messymix.com/.

21. Alexander R. Galloway and Eugene Thacker, *The Exploit: A Theory of Networks* (Minneapolis: University of Minnesota Press, 2007), 81.

22. David Evans, ed., *Appropriation* (London: MIT Press, 2009), 16–22.

Theme 5

1. Marshall McLuhan, *Understanding Media: The Extensions of Man* (Cambridge, MA: MIT Press, 1994), 69.

2. Rita Raley, *Tactical Media* (Minneapolis: University of Minnesota Press, 2009), 6

3. Leah Lievrouw, *Alternative and Activist New Media* (Cambridge, UK: Polity Press, 2011), 17.

4. Raley, *Tactical Media*, 3.

5. Lyn H. Lofland, *The Public Realm: Exploring the City's Quintessential Social Territory*, Communication and Social Order (Hawthorne, NY: Aldine de Gruyter, 1998), 13.

6. Edward A. Hirsch, "Contestational Design: Innovation for Political Activism," (Ph.D. diss., Massachusetts Institute of Technology, School of Architecture and Planning, 2008), 3, http://hdl.handle.net/1721.1/46594.

7. Nick Montfort, "Roboprotest: Artists and Engineers Make Subversive Allies.," *MIT Technology Review*, September 1, 2000, https://www.technologyreview.com/2000/09/01/236232/roboprotest/.

8. 2008 Taipei Biennial. Overview of Information available at: http://universes-in-universe.de/car/taipei/eng/2008/txt/Taipei-08-print-eng.pdf.

9. Regine Debatty, "Interview with the Institute for Applied Autonomy," *We Make Money Not Art* (blog), June 27, 2007, https://we-make-money-not-art.com/interview_with_18/.

10. Debatty, "Interview with the Institute for Applied Autonomy."

11. The Centre for Metahuman Exploration was an art, robotics and telepresence project at Carnegie Mellon University led by Peter Coppin; it included collaborators Martin C. Martin, Ryan Douglass, Brent Scott, Michael Parris and others. See Centre for Metahuman Exploration, "Project Paradise," in *Ars Electronica 98,* ed. G. Stocker, C. Schöpf, (Berlin, DE: Springer, 1998, https://doi.org/10.1007/978-3-662-38430-5_34

12. "Definitions of Terms," research-creation, Social Sciences and Humanities Research Council, Government of Canada, last modified May 4, 2021, https://www.sshrc-crsh.gc.ca/funding-financement/programs-programmes/definitions-eng.aspx#a22.

13. Debatty, "Interview with the Institute for Applied Autonomy."

14. L. A. Kauffman, *Direct Action: Protest and the Reinvention of American Radicalism* (London: Verso, 2017).

15. "StreetWriter," Institute for Applied Autonomy, July 7, 2009, archived April 24, 2016, https://web.archive.org/web/20160404203321/http://www.appliedautonomy.com/sw.html.

16. "The Grand Challenge," Defense Advanced Research Projects Agency, accessed August 25, 2021, https://www.darpa.mil/about-us/timeline/-grand-challenge-for-autonomous-vehicles.

17. Isaac Asimov, *I, Robot* (New York: Bantam Books, 1950), https://archive.org/details/irobot00asim_0.

18. Hutan Ashrafian, "AIonAI: A Humanitarian Law of Artificial Intelligence and Robotics," *Science and Engineering Ethics* 21, no. 1 (February 2015): 29–40, https://doi.org/10.1007/s11948-013-9513-9.

19. Dick Hebdige, *Subculture: The Meaning of Style* (London: Routledge, 1979); Thomas Frank, *The Conquest of Cool: Business Culture, Counterculture, and the Rise of Hip Consumerism*. (Chicago: University of Chicago Press, 1997); and Joseph Heath and Andrew Potter, *Nation of Rebels: Why Counterculture Became Consumer Culture* (New York: Harper Business, 2004).

20. Dick Hebdige and Andrew Potter, "A Critical Reframing of Subcultural Cool and Consumption" in *E-European Advances in Consumer Research*, ed. Stefania Borghini, Mary Ann McGrath, and Cele Otnes (Duluth, MN: Association for Consumer Research, 2007), 8:527–30.

21. Bill Toland, "Collaborators Miffed by Chalkbot's 'Commercialism,'" *Pittsburgh Post-Gazette*, August 15, 2009, https://www.post-gazette.com/business/tech-news/2009/08/15/Collaborators-miffed-by-Chalkbot-s-commercialism/stories/200908150140.

22. Nathan Martin, "History of Chalkbot," DeepLocal, 2009, archived August 30, 2009, https://web.archive.org/web/20090830022312/http://www.deeplocal.com/chalkbot.php.

23. "Nike Chalkbot Rips-off Streetwriter," Institute for Applied Autonomy, July 7, 2009, archived July 10, 2009, https://web.archive.org/web/20090710001351/http://www.appliedautonomy.com/chalkbot.html.

24. Martin, "History of Chalkbot."

25. "Nike Chalkbot Rips-off Streetwriter."

26. Other initiatives include the Artists-in-Labs Program, a Swiss initiative by the Zurich University of the Arts, the Institute of Cultural Studies (ICS), and the Swiss Federal Office for Culture. "Über Uns," Artists-in-Labs," accessed August 25, 2021, https://artistsinlabs.ch/.

27. Golan Levin, "New Media Artworks: Prequels to Everyday Life," *Flong Blog+News* (blog), July 19, 2009, http://www.flong.com/blog/2009/new-media-artworks-prequels-to-everyday-life/.

28. Levin, "New Media Artworks."

29. Eric Kluitenberg, *Legacies of Tactical Media: The Tactics of Occupation: From Tompkins Square to Tahrir*, Network Notebook 5 (Amsterdam: Institute of Network Cultures, 2011), 30.

Conclusions

1. Navi Radjou, Jaideep Prabhu, and Simone Ahuja, *Jugaad Innovation: A Frugal and Flexible Approach to Innovation for the 21st Century* (San Francisco: Random House India, 2012), 5.

2. Golan Levin, "New Media Artworks: Prequels to Everyday Life," *Flong Blog+News* (blog), July 19, 2009, http://www.flong.com/blog/2009/new-media-artworks-prequels-to-everyday-life/.

3. Clayton Christensen, "Disruptive Innovation," Clayton Christensen, accessed September 4, 2021, https://claytonchristensen.com/key-concepts/.

4. Lepota L. Cosmo, *Rock Love Quote—5000 Quotations On Rock N' Roll of Legends, Bands, Producers, Instrumentalists, Writers and Leading Vocals* (Lulu Press, Inc, 2016).

5. Kurt Cobain, Krist Novoselik, and Dave Grohl, Live Nirvana Interview Archive, interview by Frédéric Brebant, November 23, 1991, https://www.livenirvana.com/interviews/9111fb/index.php.

6. Nadya Tolokonnikova, *Read & Riot: A Pussy Riot Guide to Activism* (HarperCollins, 2018).

Index